SEMINARS IN MATHEMATICS

V. A. STEKLOV MATHEMATICAL INSTITUTE, LENINGRAD

ZAPISKI NAUCHNYKH SEMINAROV

LENINGRADSKOGO OTDELENIYA

MATEMATICHESKOGO INSTITUTA IM. V. A. STEKLOVA AN SSSR

ЗАПИСКИ НАУЧНЫХ СЕМИНАРОВ

ЛЕНИНГРАДСКОГО ОТДЕЛЕНИЯ

МАТЕМАТИЧЕСКОГО ИНСТИТУТА им. В.А. СТЕКЛОВА АН СССР

SEMINARS IN MATHEMATICS

V. A. Steklov Mathematical Institute, Leningrad

1	Studies in Number Theory	A. V. Malyshev, Editor
2	Convex Polyhedra with Regular Faces	V. A. Zalgaller
3	Potential Theory and Function Theory for Irregular Regions	Yu. D. Burago and V. G. Maz'ya
4	Studies in Constructive Mathematics and Mathematical Logic, Part I	A. O. Slisenko, Editor
5	Boundary Value Problems of Mathematical Physics and Related Aspects of Function Theory, Part I	V. P. Il'in, Editor
6	Kinematic Spaces	R. I. Pimenov
7	Boundary Value Problems of Mathematical Physics and Related Aspects of Function Theory, Part II	O. A. Ladyzhenskaya, Editor
8	Studies in Constructive Mathematics and Mathematical Logic, Part II	A. O. Slisenko, Editor
9	Mathematical Problems in Wave Propagation Theory	V. M. Babich, Editor
10	Isoperimetric Inequalities in the Theory of Surfaces of Bounded External Curvature	Yu. D. Burago
11	Boundary Value Problems of Mathematical Physics and Related Aspects of Function Theory, Part III	O. A. Ladyzhenskaya, Editor

SEMINARS IN MATHEMATICS
V. A. Steklov Mathematical Institute, Leningrad

Volume 9

MATHEMATICAL PROBLEMS IN WAVE PROPAGATION THEORY

Edited by V. M. Babich

Translated from Russian

CONSULTANTS BUREAU NEW YORK—LONDON 1970

The original Russian text was published in Leningrad in 1968 by offset reproduction of manuscript. The hand-written symbols have been retained in this English edition. This translation is published under an agreement with Mezhdunarodnaya Kniga, the Soviet book export agency.

Математические вопросы теории распространения волн

Library of Congress Catalog Card Number 77-103945

SBN 306-18809-0

A Division of Plenum Publishing Corporation
227 West 17th Street, New York, N. Y. 10011

United Kingdom edition published by Consultants Bureau, London
A Division of Plenum Publishing Company, Ltd.
Donington House, 30 Norfolk Street, London, W.C. 2, England

Printed in the United States of America

PREFACE

The papers comprising this collection are directly or indirectly related to an important branch of mathematical physics – the mathematical theory of wave propagation and diffraction.

The paper by V. M. Babich is concerned with the application of the parabolic-equation method (of Academician V. A. Fok and M. A. Leontovich) to the problem of the asymptotic behavior of eigenfunctions concentrated in a neighborhood of a closed geodesic in a Riemannian space. The techniques used in this paper have been found useful in solving certain problems in the theory of open resonators.

The topic of G. P. Astrakhantsev's paper is similar to that of the paper by V. M. Babich. Here also the parabolic-equation method is used to find the asymptotic solution of the elasticity equations which describes Love waves concentrated in a neighborhood of some surface ray.

The paper of T. F. Pankratova is concerned with finding the asymptotic behavior of the eigenfunctions of the Laplace operator from the exact solution for the surface of a triaxial ellipsoid and the region exterior to it.

The first three papers of B. G. Nikolaev are somewhat apart from the central theme of the collection; they treat the integral transforms with respect to associated Legendre functions of first kind and their applications. Examples of such applications are the use of this transform for the solution of integral equations with symmetric kernels and for the solution of certain problems in the theory of electrical prospecting.

The two papers by B. G. Nikolaev are devoted to describing one of the possible approaches to the solution of stationary problems related to the diffraction of waves by angular plane regions. Examples considered are the Neumann problem for the exterior of a wedge-shaped region and the Neumann problem for a half space divided by an inclined boundary into two angular regions with different wave-propagation speeds.

The paper by A. I. Lanin is a continuation of the joint work of V. S. Buldyrev and A. I. Lanin on interference waves in diffraction problems for the cylinder and sphere. The intensities of interference-type waves and of noninterfering waves of geometrical optics are computed on the basis of tabulated results for the special functions $G_M(\gamma)$ and $\Gamma_M(\gamma)$, which describe the interference fields. The behavior of the field in a neighborhood of a limiting ray for the sphere and planar case is compared.

In the paper "Calculation of the wave fields for multiple waves near the points of origin" by N. S. Smirnova the principal parts of the displacement fields are studied for reflected waves and for multiple head waves of arbitrary type which arise in a system of n plane-parallel elastic layers. Attention is focused on obtaining computational formulas which describe the displacement fields near the point of origin of the head wave. A number of particular examples are considered.

CONTENTS

SHARPLY DIRECTED PROPAGATION OF LOVE-TYPE SURFACE WAVES

G. P. Astrakhantsev

In this paper we construct displacement vectors $\vec{u} = exp[-i\omega(t-\tau)]\vec{\varphi}(\tau, \alpha, \nu, \omega)$, which asymptotically (for $\omega \to \infty$) satisfy the equations of the theory of elasticity and the condition of absence of stresses on the boundary S of an arbitrary elastic body and which are concentrated in a neighborhood of a ray $\mathcal{L}$ on S. The vector $\vec{\varphi}$ is constructed by the parabolic-equation method.

§1. Derivation of the Parabolic Equations

We consider an elastic body with Lamé parameters λ, μ and density ρ, which depend on the coordinates (x, y, z). A ray is understood to mean an extremal of the integral

$$\int_{M_1}^{M_2} \frac{ds}{b(x,y,z)}, \qquad b^2 = \frac{\mu}{\rho}. \tag{1}$$

In addition to the Cartesian coordinate system (x, y, z), we shall consider a system of curvilinear coordinates (q^1, q^2, q^3) in a neighborhood of the ray $\mathcal{L}$ on S. To points on $\mathcal{L}$ we assign the coordinates

$$q^1 = \tau(x, y, z) = \int_{M_0}^{M} \frac{ds}{b}$$

(M_0 and M are points on $\mathcal{L}$, and the integration is along $\mathcal{L}$). We consider the extremals of the integral (1) on S which are orthogonal to $\mathcal{L}$; to points on such rays we assign the coordinates $q^2 = \alpha(x, y, z) = \int_{\tau}^{M} \frac{ds}{b}$.
The parameter $q^3 \equiv \nu$ is the distance along the normal to the given surface, whereby points of the elastic body correspond to values $\nu \geqslant 0$.

We assume that volume forces are absent and that the surface S is free of stresses. In the curvilinear coordinate system (q^1, q^2, q^3) let the element of length be $ds = g_{ij}\, dq^i dq^j$, where the g_{ij} are the coefficients of the metric tensor. For small vector displacements we have in the coordinate system (q^1, q^2, q^3) the system of equations:

$$\sigma_{i'j'}\, g_s^{ii'} g^{jj'} \sqrt{g}\, \frac{\partial g_{ij}}{\partial q^s} + 2\frac{\partial}{\partial q^j}(\sigma_{sj}\, g^{jj'}\sqrt{g}) - 2\rho g_{si}\sqrt{g}\,\frac{\partial^2 u^i}{\partial t^2} = 0, \tag{2}$$

where the index varies from 1 to 3, $g = det\,\|g_{ij}\|$, $\|g^{ij}\| = \|g_{\alpha\beta}\|^{-1}$, and the σ_{ij} are the components of the stress tensor:

$$\sigma_{ij}(u^s) = \frac{\lambda}{\sqrt{g}} \frac{\partial}{\partial q^s}(u^s\sqrt{g})\, g_{ij} + \mu\left[\frac{\partial g_{ij}}{\partial q^s} u^s + g_{si}\frac{\partial u^s}{\partial q^j} + g_{is}\frac{\partial u^s}{\partial q^i}\right].$$

The components of the vector $\vec{u}$ in the coordinate system (q^1, q^2, q^3) are related to the projections of the vector $\vec{u} = (u_1\ u_2\ u_3)$ in the Cartesian coordinate system $(x, y, z) = (x_1, x_2, x_3)$ by the formula

$$u_i = u^j \frac{\partial x_i}{\partial q^j}. \tag{3}$$

The vector $\vec{u}$ is such that on the boundary there are no stresses, i.e.,

$$\sigma_{i3}\big|_{\nu=0} = 0. \tag{4}$$

If $\mathcal{L}$ is an extremal of the integral (1), then the first quadratic form on S in a neighborhood of $\mathcal{L}$ has the form:

$$\widetilde{ds}^2 = b^2(\tau, \alpha, 0)\, d\alpha^2 + E(\tau, \alpha)\, d\tau^2,$$

where moreover

$$E(\tau, \alpha)\big|_{\alpha=0} = b^2(\tau), \quad E_\alpha(\tau, \alpha)\big|_{\alpha=0} = (b^2)_\alpha.$$

The coefficients of the metric tensor g_{ij} with an accuracy up to $O(\nu + \alpha^2)$ are given by

$$\begin{aligned} g_{11} &= -2\nu L + b^2(\tau) + (b^2)_\alpha \alpha + E_{\alpha\alpha}\frac{\alpha^2}{2}, \\ g_{22} &= -2\nu N + b^2(\tau) + (b^2)_\alpha \alpha + (b^2)_{\alpha\alpha}\frac{\alpha^2}{2}, \\ g_{12} &= -\nu M, \quad g_{33} = 1, \quad g_{31} = g_{32} = 0, \end{aligned} \tag{5}$$

where L, M, and N are the coefficients of the second quadratic form on S.

Suppose that the displacement vector has the form

$$\begin{aligned} \vec{u} &= \exp\left[-i\omega(t-\tau)\right]\vec{\varphi}, \\ \vec{\varphi}(\tau, \alpha, \nu) &= (\varphi^\tau, \varphi^\alpha, \varphi^\nu). \end{aligned} \tag{6}$$

We seek a vector $\vec{\varphi}$, having the properties of Love waves, that is, the components φ^τ and φ^ν are small compared with φ^α, and φ^α is appreciably different from zero only in a neighborhood of the ray $\mathcal{L}$. As concerns φ^i, we shall assume that

$$\frac{\partial^{k+\ell+m}}{\partial\tau^k \partial\alpha^\ell \partial\nu^m}\varphi^i = O(\omega^{\frac{k}{3}} . \omega^{\frac{\ell}{2}} . \omega^{\frac{2m}{3}}) \max|\varphi^i|.$$

We substitute (6) into equations (2) and (4). Retaining only those terms of the equations which near $\mathcal{L}$ are order $O(\omega)$ or higher with respect to the parameter ω $(\omega \to \infty)$, we obtain equations and corresponding boundary conditions for φ^i.

Let

$$\varphi^\tau = \omega^{-\frac{1}{2}} A_1 + \omega^{-\frac{2}{3}} A_2 + \omega^{-1} A_4, \quad \varphi^\nu = \omega^{-\frac{1}{3}} A_3, \quad \varphi^\alpha = w$$

where $A_i = O(\varphi^\alpha)$.

If

$$A_1 = i\omega^{-\frac{1}{3}} \frac{\partial W}{\partial \alpha}, \quad A_2 = i\omega^{-\frac{2}{3}} \frac{\partial A_3}{\partial \nu}, \tag{7}$$

$$A_4 = i\,\frac{\left[\frac{\lambda}{\sqrt{g}}(\sqrt{g})_\alpha + \mu\frac{\partial g_{11}}{\partial \alpha}g^{11} + \frac{1}{\sqrt{g}}\frac{\partial}{\partial \alpha}(\mu\sqrt{g})\right]}{(\lambda+\mu)} \cdot W, \tag{8}$$

$$2i\omega^{\frac{2}{3}}\mu\frac{\partial A_3}{\partial \tau} + \omega^{-\frac{1}{3}}\mu b^2\frac{\partial^2 A_3}{\partial \nu^2} - 2\nu\mu\omega^{\frac{5}{3}}\left[\kappa + (\ln b)_\nu\right]A_3 = i\omega M\mu W, \tag{9}$$

$$2i\omega\mu b^4 W_\tau + \omega\mu b^6 W_{\nu\nu} - 2\nu\mu\omega^2 b^4\left[\kappa + (\ln b)_\nu\right]W + b^2\frac{\partial(\mu b^2)}{\partial \tau}i\omega W + \mu b^4 W_{\alpha\alpha} + \omega^2\mu\alpha^2 R(\tau)W = 0, \tag{10}$$

$$A_3\Big|_{\nu=0} = 0, \quad \frac{\partial W}{\partial \nu}\Big|_{\nu=0} = 0, \tag{11}$$

then equations (2) are satisfied with an accuracy up to $O(\omega)$ [the right sides have order $O(\omega^{\frac{5}{6}})$], and the boundary conditions (4) are satisfied with an accuracy up to $O(\omega^{\frac{2}{3}})$.

Here

$$R(\tau) \equiv \frac{1}{2}\left(\frac{E}{b^2}\right)_{\alpha\alpha}\Big|_{\alpha=0} \tag{12}$$

and κ is the curvature of the normal section of the surface in the direction of the ray $\mathcal{L}$. We shall assume that on the ray $\mathcal{L}$ the following condition is satisfied:

$$\kappa + (\ln b)_\nu > 0. \tag{13}$$

We will solve equations (9) and (10) by techniques used in [1].

§2. Integration of Equations (9) and (10)

We introduce the variable ξ,

$$\xi = \omega^{\frac{2}{3}}\chi(\tau)\nu. \tag{14}$$

Equation (10) is then written as follows:

$$2i\omega\mu b^4 w_\tau + \mu b^6\chi^2\omega^{\frac{4}{3}}\left[w_{\xi\xi} - \xi w\right] + 2i\omega\mu b^4\frac{\chi_\tau}{\chi}\xi w_\xi + \mu b^4 w_{\alpha\alpha} + \omega^2 b^4\mu\alpha^2 R w + b^2\frac{\partial}{\partial \tau}(\mu b^2)i\omega w = 0,$$

where

$$\chi(\tau) = 2^{\frac{1}{3}}b^{-\frac{2}{3}}\left(\kappa + (\ln b)_\nu\right)^{\frac{1}{3}}. \tag{15}$$

The function $\chi(\tau)$ is found from the condition that the terms in square brackets have the same coefficients of τ. We now eliminate the term $2i\omega\mu b^4\frac{\chi_\tau}{\chi}\xi w_\xi$, by putting

$$w = V\exp\left[i\beta(\tau)\omega^{-\frac{1}{3}}\xi^2\right].$$

Determining β from the condition

$$\frac{\chi_\tau}{\chi} + 2\beta b^2\chi = 0, \tag{16}$$

we arrive at the equation for V:

$$2i\omega\mu b^4 V_\tau + \mu b^6 \chi^2 \omega^{\frac{4}{3}}(V_{\xi\xi} - \xi V) + 2i\omega\mu b^6 \chi^2_\beta V + b^2(\mu b^2)_\tau i\omega V + \mu b^4 V_{\alpha\alpha} + \omega^2 b^4 \mu \alpha^2 R V = 0.$$

Let $V = Ai(\xi - t)\tilde{V}(\tau, \alpha, \omega)$, where Ai is some solution of the Airy equation, and t is a constant which is interpreted as a separation constant. Condition (11) and $w \underset{\nu\to+\infty}{\longrightarrow} 0$ imply that Ai should be taken to be the Airy function $V(t)$ such that

$$V(t) \sim \frac{1}{2} t^{-\frac{1}{4}} exp[-\frac{2}{3} t^{\frac{3}{2}}],$$

and $t = t'_\kappa$, where t'_κ $(\kappa = 1, 2, \ldots)$ are roots of the equation

$$V'(-t) = 0. \tag{17}$$

We put

$$\tilde{V} = u_1(\tau, \alpha, \omega)\sqrt{\frac{\chi}{\mu b^2}}\, exp[-\frac{i}{2}\int_0^\tau \omega^{\frac{1}{3}} b^2 \chi^2 t'_\kappa d\tau].$$

Assuming that $\frac{\partial u_1}{\partial \tau} = O(1)$ we arrive at the following equation for u_1:

$$2i\omega u_{1\tau} + u_{1\alpha\alpha} + \omega^2 \alpha^2 R u_1 = 0. \tag{18}$$

Similarly,

$$B = exp[\frac{i}{2}\int_0^\tau \omega^{\frac{1}{3}} b^2 \chi^2 t'_\kappa d\tau] A_s, \tag{19}$$

we obtain the equation

$$B_{\xi\xi} - (\xi - t'_\kappa) B = (\chi b)^{-2} M exp[\frac{i}{2}\int_0^\tau \omega^{\frac{1}{3}} b^2 \chi^2 t'_\kappa d\tau] w, \tag{20}$$

with boundary conditions

$$B|_{\xi=0} = 0, \qquad B \underset{\xi\to+\infty}{\longrightarrow} 0. \tag{21}$$

The homogeneous adjoint equation for the Sturm–Liouville problem (20), (21) has only zero solutions satisfying conditions (21). This implies that problem (20), (21) is solvable for any right-hand side.

We seek the function u_1 in the form

$$u_1 = \tilde{u}\, exp[i\omega\delta(\tau)\alpha^2] = \tilde{u}\, exp[i\delta\gamma^2(\tau)\eta],$$

where

$$\alpha = \omega^{\frac{1}{2}} \gamma(\tau)\eta. \tag{22}$$

For separation of the variables it is sufficient that γ and δ satisfy the conditions:

$$2\delta - \frac{\gamma'}{\gamma} = 0, \quad 4\delta^2 + 2\delta' - R = \frac{c}{\gamma^4}. \tag{23}$$

Eliminating δ from (23), we obtain

$$\gamma'' - R(\tau)\gamma = \frac{c}{\gamma^3}. \tag{24}$$

Solutions of equation (24) must not have zeros (otherwise the function $\tilde{u}$ will have singularities).

The equation for $\tilde{u}$ is rewritten as follows:

$$\gamma^{-2}(\tilde{u}_{\eta\eta} - c\eta^2\tilde{u}) + 2i(\tilde{u}_\tau + \delta(\tau)\tilde{u}) = 0$$

We seek a solution in the form $\mathcal{D}(\eta)\ S(\tau)$. Separating variables, we obtain

$$\mathcal{D}_{\eta\eta} + (\ell - c\eta^2)\mathcal{D} = 0, \tag{25}$$

$$S_\tau + (\delta - \frac{\ell}{2i\gamma^2})S = 0. \tag{26}$$

For $c < 0$ equation (25) has no solutions which tend to zero for $\eta \to \infty$. We assume that $c > 0$, and put $c = 1$. For $c > 0$ the solutions of equation (24) have no zeros (see [1]). Solutions $\mathcal{D}(\eta)$, which tend to zero for $\eta \to \infty$, exist for $\ell = 2m + 1$, $(m = 0, 1, 2, \ldots)$. In this case

$$\mathcal{D}_m(\eta) = \exp\left[-\frac{\eta^2}{2}\right] H_m(\eta),$$

where H_m is a Hermite polynomial.

As a result, for the displacement vector $\vec{u}$ we have obtained the asymptotic expressions:

$$u^1 = \exp[-i\omega(t-\tau)] \cdot [i\omega^{-1} W_\alpha + i\omega^{-\frac{4}{3}} \frac{\partial A_3}{\partial \nu} + \omega^{-1} A_4]$$

$$u^2 = \exp[-i\omega(t-\tau)] W,$$

$$u^3 = \exp[-i\omega(t-\tau)] \cdot [\omega^{-\frac{1}{3}} A_3],$$

where W has the form

$$W = \sqrt{\frac{\chi}{\mu b^2 \gamma}}\ \vartheta(\tau - t_k')\exp\left[i\omega\frac{\gamma'}{2\gamma}\alpha^2 - \frac{\eta^2}{2}\right] \cdot \exp\left[-\frac{i}{2}\int_0^\tau (\omega^{\frac{1}{3}} b^2 \chi^2 t_k' + \frac{2m+1}{\gamma^2})\,d\tau\right] H_m(\eta),$$

$$m = 0, 1, 2, \ldots, \quad k = 1, 2, 3, \ldots$$

The functions A_3 and A_4 are defined by equations (20) and (8).

In conclusion, the author would like to thank V. M. Babich for suggesting this problem and for providing a great deal of assistance.

LITERATURE CITED

1. Babich, V. M., and Lazutkin, V. F. "Eigenfunctions concentrated near a closed geodesic," in: Topics in Mathematical Physics, Vol. 2, M. Sh. Birman, ed., Consultants Bureau, New York, (1968).
2. Babich, V. M., and Molotkov, I. A., "Propagation of Love waves in an elastic half space which is inhomogeneous with respect to two coordinates," Izv. Akad. Nauk SSSR, Fizika Zemli, Vol. 6 (1966).

EIGENFUNCTIONS CONCENTRATED IN A NEIGHBORHOOD OF A CLOSED GEODESIC

V. M. Babich

§1. Introduction

The asymptotic behavior of the eigenfunctions for the triaxial ellipsoid

$$\frac{x^2}{a^2}+\frac{y^2}{b^2}+\frac{z^2}{c^2}=1, \quad a>b>c. \tag{1.1}$$

has been studied in the papers of V. P. Bykov [1] and L. A. Vainshtein [2].

It follows from their results that there exist subsequences of eigenfunctions which are "concentrated"* in a neighborhood of the ellipses

$$\frac{x^2}{a^2}+\frac{y^2}{b^2}=1, \quad z=0, \tag{1.2}$$

$$\frac{y^2}{b^2}+\frac{z^2}{c^2}=1, \quad x=0. \tag{1.3}$$

The ellipses (1.2) and (1.3) are obtained as the intersections of the ellipsoid (1.1) with the coordinate planes $x=0$ and $z=0$. It is interesting to note that there is no subsequence of eigenfunctions which is concentrated in a neighborhood of the ellipse

$$\frac{x^2}{a^2}+\frac{z^2}{c^2}=1, \quad y=0 \tag{1.4}$$

The ellipses (1.2), (1.3), and (1.4) are closed geodesics on the surface of the ellipsoid.

Let Ω be an arbitrary finite three-dimensional region which is bounded by a sufficiently smooth surface S. It is natural to ask the following question: when can eigenfunctions be concentrated in a neighborhood of a closed geodesic on S?

A neighborhood of a closed geodesic in which an eigenfunction is "concentrated" is a typical "boundary layer"; therefore, in order to find an asymptotic expression for such eigenfunctions, it is natural to try to apply the corresponding technique, i.e., the classical parabolic-equation method (cf. [3], [4]).

The attempt to find the asymptotic behavior of an eigenfunction in the first approximation is successful if and only if the corresponding closed geodesic is stable. Stable geodesics have the property that geodesics near them "wind a sufficiently long time" in a neighborhood of them. The "largest"

*That is, they are appreciably different from zero only near the ellipses (1.2) and (1.3) and fall off very rapidly away from these ellipses (for more details, see [1], [2], and also § 3 and the following sections of the present paper).

ellipse (1.2) and the "smallest" ellipse (1.3) are stable, while the ellipse (1.4) of "intermediate size" is unstable in the sense of this definition (cf. [5] and [6]).

In the present paper, using the parabolic-equation method, we find the asymptotic behavior of eigenfunctions of the Laplace operator concentrated in a neighborhood of a closed geodesic in an $(m+1)$-dimensional Riemannian space.

In contrast to the particular case $m=1$ considered in [6], we have not succeeded in separating the variables in the parabolic equation of our problem for $m>1$ (in spite of a number of determined efforts). It was possible to integrate the parabolic equation by quite another approach. All of our constructions are possible only for the case of a stable geodesic (cf. § 4).

In § 4 we prove a number of propositions concerning the solutions of the Jacobi equation for the geodesic in question. All these results can be obtained directly by carrying over the proofs for the corresponding results in the case of linear Hamiltonian systems with periodic coefficients.

We have preferred to present detailed proofs (which differ from the "classical" proofs) in order to have the results in a convenient form and also in order not to inconvenience the reader with references to literature which is not readily accessible.

The problem considered in the present paper is of interest not only "in itself," but also as a "model" for many problems related to open resonators and the theory of sharply directed propagation of short waves. Using the technique suggested in the present paper, M. M. Popov recently found the asymptotic behavior of the proper modes of an open resonator with nonparallel mirrors.

§ 2. The Formulation of the Problem and the Ray Method in a Riemannian Space

Let some region of an $(m+1)$-dimensional Riemannian space be such that the system of coordinates $(q^0, q^1, \ldots q^m)$ is regular. The element of length then has the following form:

$$d\tau = \sqrt{g_{ij}\, dq^i dq^j} \tag{2.1}$$

(summation over repeated indices is understood), where

$$g_{ij}(q^0, \ldots q^m)$$

are the components of the metric tensor, which we shall assume to be sufficiently smooth functions. The Laplace operator in a Riemannian space is the differential operator

$$\Delta_2 u = \frac{1}{\sqrt{g}} \frac{\partial}{\partial q^i} \left(g^{ij} \sqrt{g} \frac{\partial u}{\partial q^j} \right), \tag{2.2}$$

where $\| g^{ij} \|$ is the inverse of the matrix $\| g_{ij} \|$, $g = \det \| g_{ij} \|$.

The Laplace operator is invariant under a change of the variables q^i. The analogue of the Helmholtz equation in a Riemannian space is the equation

$$\Delta_2 u + k^2 u = 0. \tag{2.3}$$

Just as in the case of the classical Helmholtz equation, it is natural to try to construct its solution formally in a manner analogous to the usual ray expansion (see, for example, [7]):

$$\left. u = e^{ik\tau(q)} \sum_{s=0}^{\infty} \frac{u_s(q)}{(-ik)^{s+\gamma}}, \quad \begin{array}{l} q = (q^0, \ldots q^m); \\ \gamma = \mathrm{const}; \; (-i)^{s+\gamma} = e^{-\frac{\pi}{2} i (s+\gamma)} \end{array} \right\}. \tag{2.4}$$

τ and u_s are unknown functions.

Substituting expansion (2.4) into equation (2.3) and equating the coefficients of successive powers of $\frac{1}{k}$ to zero, we obtain

$$g^{ij}\frac{\partial\tau}{\partial q^i}\frac{\partial\tau}{\partial q^j}=1\,, \tag{2.5}$$

$$2g^{ij}\frac{\partial\tau}{\partial q^i}\frac{\partial u_s}{\partial q^j}+u_s\Delta_2\tau=\Delta u_{s-1}\,, \qquad s=0,1,2,\dots\,;\quad u_{-1}\equiv 0 \tag{2.6}$$

Equation (2.5) is the analogue of the classical eikonal equation $(grad\ \tau)^2=1$. Indeed, it is easily seen that equation (2.4) has the same form in any system of coordinates. If, in particular, we choose a system of coordinates such that at the point q_0 the matrix $\|g_{ij}\|$ is the identity matrix, then at this point equation (2.5) becomes

$$\sum_{j=0}^{m}\left(\frac{\partial\tau}{\partial x_j}\right)^2=1\,,$$

where the x_j refer to the system of coordinates in question. The derivative $\frac{\partial\tau}{\partial x_j}$ obviously coincides with the derivatibes of τ in the direction of the x_j coordinate axis along an arc corresponding to a coordinate line. The coordinate system x_j at the point q^0 is evidently orthogonal. Hence, the left side of equation (2.5) [just as in the classical case of the equation $(grad\ \tau)^2=1$] is the sum of the squares of the derivatives of τ in $m+1$ mutually perpendicular directions.

It is not difficult to construct a solution of equation (2.5) as an extremal of the integral

$$\int d\sigma=\int\sqrt{g_{ij}\,dq^i dq^j} \tag{2.7}$$

It can be shown that the function

$$\tau(M)=\int_{M_0}^{M}d\sigma=\int_{M_0}^{M}\sqrt{g_{ij}\,dq^i dq^j} \tag{2.8}$$

is a solution of the eikonal equation (2.5) [the point M_0 is fixed, and the integral is taken along all extremals of the integral (2.7) passing through M_0]. Similarly, the function (2.8) is a solution of equation (2.6) if the point M ranges over a sufficiently smooth m-dimensional hyperplane; the integral (2.8) is taken along the extremals of the integral (2.7), which intersect this hyperplane orthogonally at the point M_0.

Recalling how the solution of the eikonal equation in the classical case is constructed "by rays" (cf. [7]), we may say that in the case under consideration the extremals of the integral (2.7) play the role of rays. The extremals of the integral (2.7) are called geodesic lines of the Riemannian space, and they play a fundamental role in the theory of such spaces. For brevity, we shall refer to them simply as geodesics.

We shall not consider equations (2.8) at length, because of their complete similarity to the recurrent system of the ray method [7].

We shall be interested in the eigenfunctions concentrated in a neighborhood of some closed geodesic ℓ; more precisely, we shall be interested in the formal solutions for "large" κ of equation (2.3) which are of order $O(1)$ in a neighborhood of ℓ and tend rapidly to zero away from ℓ. A similar two-dimensional $(m+1=2)$ is considered in [6]. If the problem concerning the eigenfunctions is considered in a strict formulation, that is, if we consider, for example, that our Riemannian space is a compact manifold (without boundary) and that an eigenfunction is a nonzero solution of equation (2.3), then there is reason to believe that the sequence of "eigenfunctions" and "eigenvalues" which we have found in a formal way actually gives the asymptotic behavior of a certain subsequence of "genuine" eigenvalues and eigenfunctions in a broad class of cases.

§ 3. The Parabolic Equation for the Problem and Related Constructions

Let ℓ be a closed geodesic in an $(m+1)$-dimensional Riemannian space.

We shall describe the coordinate system in which it is convenient to carry out the subsequent constructions. We shall characterize points on ℓ by the arc length s (arc length is understood in the sense of the Riemannian space in question), which is measured along ℓ in either direction from a fixed point. For convenience, we shall assume that $-\infty < s < +\infty$.

Let s be some point on ℓ. We consider all geodesics through s which are orthogonal to ℓ at s. In the submanifold F_s, formed from these geodesics, we introduce so-called normal Riemannian coordinates: let $e_1, e_2, \ldots e_m$ be some orthonormal system of vectors which are orthogonal to ℓ at the point s.* In order to locate an arbitrary point M in F_s, it is sufficient to give the geodesic passing through s and M (and orthogonal to ℓ) and the arc length $|OM|$ along this geodesic; for this, it is in turn sufficient to give the m numbers

$$y_j = |OM| \cos \widehat{i_0 e_j}$$

(i_0 is a unit vector at the point O of the geodesic OM, which is "directed from O to M").

The quantities y_j and s are called Riemannian normal coordinates (cf. [8]). It is well known that in a neighborhood of the point $\{y_j = 0\}$ $j = 1, 2, \ldots m$ this system of coordinates is regular. By translating parallel to $\vec{e}_j$ $j = 1, 2, \ldots m$ ["parallel" in the sense of parallel along a curve ℓ in Riemannian space (cf. [8])] we can introduce Riemannian coordinates in any submanifold $F_{s'}$ $s' \in \ell$. Because of the fact that the scalar product of any pair of vectors does not change under parallel translation, the vectors $\vec{e}_0, \vec{e}_1, \ldots \vec{e}_m$, where $\vec{e}_0$ is the unit vector tangent to ℓ, always form an orthonormal system of vectors. (We recall that in translating the unit vector $\vec{e}_0$ tangent to ℓ at the point s parallel along ℓ, we again obtain at each point of ℓ a unit vector tangent to ℓ. This follows from that fact that ℓ is a geodesic.)

Thus, in the neighborhood of any segment of ℓ we have a well-defined, regular coordinate system $(s, y_1, \ldots y_m)$ which is orthonormal on ℓ. The fact that the system of coordinates $(s, y_1, \ldots y_m)$ is orthonormal implies that on ℓ

$$g_{ij} = \delta_{ij} = \begin{cases} 1, & i = j \\ 0, & i \neq j \end{cases} \qquad i, j = 0, 1, \ldots m \tag{3.1}$$

(the g_{ij} are the components of the metric tensor in the coordinates y^j, where $s = y^0$, $y_j = y^j$, $j > 0$).

In general, there is no such system of coordinates in an entire neighborhood of ℓ, since in translating the vectors $\vec{e}_1, \ldots \vec{e}_m$ parallel along ℓ we return finally to the starting point; the vectors $\vec{e}_1, \ldots \vec{e}_m$ hereby go over into vectors $\vec{e}'_1, \ldots \vec{e}'_m$. The system of vectors $\vec{e}_1, \ldots \vec{e}_m$ will, in general, not coincide with the vectors $\vec{e}'_1, \ldots \vec{e}'_m$ if $m > 1$.

In the two-dimensional case, $m = 1$ and there are only two possibilities: 1) $\vec{e}_1 = \vec{e}'_1$ (which happens if and only if the geodesic ℓ lies reversibly in the two-dimensional Riemannian space in question) and 2) $\vec{e}'_1 = -\vec{e}_1$. The impossibility of introducing coordinates $(s, y_1, \ldots y_m)$ in an entire neighborhood of ℓ causes no difficulty, since if some construction is not to be carried out "locally," then the corresponding equations will be written in an invariant, coordinate-free form.

We now derive certain properties of the coordinates $s, y_1, \ldots y_m$ $(\equiv y^0, y^1, \ldots y^m)$.

The set of points $\{s, y_1\gamma, \ldots y_m\gamma;\ 0 \leq \gamma \leq 1\}$ represents a segment of a geodesic orthogonal to ℓ at the point s; therefore, the following equations must be satisfied (cf. [8]):

$$\frac{d^2 y^j}{d\gamma^2} = -\Gamma^j_{ab} \frac{dy^a}{d\gamma}\frac{dy^b}{d\gamma}; \quad \Gamma^j_{ab} = \frac{1}{2} g^{jc}\left(\frac{\partial g_{ac}}{\partial y^b} + \frac{\partial g_{bc}}{\partial y^a} - \frac{\partial g_{ab}}{\partial y^c}\right), \quad (y^0, y^1, \ldots, y^m) = (0, y_1\gamma, \ldots y_m\gamma); \ 0 \leq \gamma \leq 1$$

*We recall that the Riemannian space considered is $(m+1)$-dimensional.

(Γ^{i}_{ab} is the Christoffel symbol) or

$$\sum_{a,b=1}^{m} \Gamma^{i}_{ab} y_a y_b \equiv 0,$$

whence on ℓ

$$\Gamma^{j}_{ab}\Big|_{\ell} = 0, \quad j, a, b = 1, 2, \ldots m. \tag{3.2}$$

Translating the vectors $e_j = (0,..,1,..,0)$ parallel along ℓ we always obtain a vector with the same components. Using the classical formula for the infinitesimal increment $\delta v^{\varkappa}$ of the components of the vector $\{v^j\}$, translated parallel along the infinitesimal vector $\{\delta y^i\}$ (cf. [8]),

$$\delta v^{\kappa} = -\Gamma^{\kappa}_{ij} \delta y^i v^j,$$

we obtain in addition to equations (3.2)

$$\Gamma^{\varkappa}_{oj}\Big|_{\ell} = 0, \quad \kappa, j = 0, 1, \ldots m. \tag{3.3}$$

Using the well-known formula

$$\frac{\partial g_{i\kappa}}{\partial y^j} = g_{\kappa a} \Gamma^{a}_{ij} + g_{ia} \Gamma^{a}_{\kappa j}$$

and equations (3.2) and (3.3), we obtain

$$\frac{\partial g_{i\kappa}}{\partial y_j}\Big|_{\ell} = 0, \quad i, j, \kappa = 0, 1, \ldots m, \quad \frac{\partial}{\partial y^0} = \frac{\partial}{\partial s}. \tag{3.4}$$

The classical formula for the components of the curvature tensor $R_{ij\kappa r}$ (cf. [8]),

$$R_{ij\kappa r} = \frac{1}{2}\left(-\frac{\partial^2 g_{ri}}{\partial y^j \partial y^\kappa} - \frac{\partial^2 g_{i\kappa}}{\partial y^j \partial y^r} + \frac{\partial^2 g_{rj}}{\partial y^i \partial y^\kappa} + \frac{\partial^2 g_{j\kappa}}{\partial y^i \partial y^r}\right) + g_{ab}\left(\Gamma^{b}_{ir} \Gamma^{a}_{j\kappa} - \Gamma^{b}_{jr} \Gamma^{a}_{i\kappa}\right),$$

gives

$$R_{oioj} = \frac{1}{2} \frac{\partial^2 g_{oo}}{\partial y_i \partial y_j}\Big|_{\ell}, \quad i, j = 1, 2, \ldots m,$$

that is,

$$g_{oo} = 1 - \sum_{i,j=1}^{m} K_{ij}(s) y_i y_j + O\left[\left(\sum_{j=1}^{m} y_j^2\right)^{\frac{3}{2}}\right];$$
$$K_{ij}(s) = K_{ji}(s) = R_{oioj}\Big|_{\ell}. \tag{3.5}$$

We now proceed to the derivation of the parabolic equation of the problem.

We write the equation for the eigenfunctions of $(\Delta + \kappa^2)u = 0$ in the coordinates $s, y_1, \ldots y_m$:

$$\Delta_2 u + \kappa^2 u \equiv \frac{1}{\sqrt{g}} \frac{\partial}{\partial y^i}\left(\sqrt{g}\, g^{ij} \frac{\partial u}{\partial y^j}\right) + \kappa^2 u = 0,$$
$$(s, y_1, y_2, \ldots y_m) = (y^0, \ldots y^m), \quad \|g^{ij}\| = \|g_{ij}\|^{-1}, \quad g = \det\|g_{ij}\|. \tag{3.6}$$

We shall seek a solution of the form

$$u = e^{i\kappa s}\,\mathcal{U}\,, \tag{3.7}$$

where $\mathcal{U}$ is a new unknown function (the "attenuation function" in the language of V. A. Fok).

In deriving the parabolic equation for $\mathcal{U}$ we shall assume that κ is "large" and that

$$\mathcal{U} = O(1), \quad \frac{\partial^{\tau_1+\tau_2+\tau_3}\mathcal{U}}{\partial s^{\tau_1}\partial y_i^{\tau_2}\partial y_j^{\tau_3}} = O(\kappa^{\frac{\tau_2+\tau_3}{2}}), \quad |y_j| = O(\kappa^{-\frac{1}{2}}). \tag{3.8}$$

The assumptions (3.8) are in agreement with the formulas which will eventually be obtained for $\mathcal{U}$.

In deriving the parabolic equation for $\mathcal{U}$ the relations (3.1) and (3.4) will play an important role. From them it follows easily that

$$\begin{gathered}\sqrt{g} = \sqrt{1+O(|y|^2)} = 1+O(|y|^2); \quad \frac{1}{\sqrt{g}} = 1+O(|y|^2); \\ \|g^{ij}\| = \|g_{ij}\|^{-1} = \|I+g_{ij}-I\| = I-(\|g_{ij}\|-I)+(\|g_{ij}\|-I)^2+\ldots, \quad |y|^2 = \sum_{j=1}^{m} y_j^2 .\end{gathered} \tag{3.9}$$

I here denotes the identity matrix.

From the last expression for $\|g^{ij}\|$ and formulas (3.4) and (3.5) it follows that

$$\|g^{ij}\| = 1+O(|y|^2), \quad g^{00} = 1+\sum_{i,j=1}^{m} K_{ij}(s)\,y_i\,y_j + O(|y|^3). \tag{3.10}$$

If we substitute expression (3.7) into equation (3.6), use formulas (3.8)-(3.10), and neglect terms of order $O(\sqrt{\kappa})$ and lower, we obtain the following equation for $\mathcal{U}$:

$$2i\kappa\frac{\partial\mathcal{U}}{\partial s} + \sum_{j=1}^{m}\frac{\partial^2\mathcal{U}}{\partial y_j^2} - \kappa^2\sum_{j,h=1}^{m} K_{jh}(s)\,y_j\,y_h\,\mathcal{U} = 0 \tag{3.11}$$

[it is not difficult to see that each term in equation (3.11) has order $O(\kappa)$]. We divide both sides of equation (3.11) by κ and introduce the new variables

$$\mu_j = \sqrt{\kappa}\,y_j\,. \tag{3.12}$$

Equation (3.11) then assumes the form

$$\mathcal{L}\,\mathcal{U} = 2i\frac{\partial\mathcal{U}}{\partial s} + \sum_{j=1}^{m}\frac{\partial^2\mathcal{U}}{\partial\mu_j^2} - \sum_{j,h=1}^{m} K_{jh}\,\mu_j\,\mu_h\,\mathcal{U} = 0 \tag{3.13}$$

Equation (3.13) is sometimes written in a more invariant form.

If we assign to each point s of the geodesic ℓ an m-dimensional hyperplane Φ_s, normal to ℓ and passing through s, we obtain a geometrical object Ξ called the normal bundle.*

Equation (3.13) can be considered as an equation on Ξ: the unknown function $\mathcal{U}$ is a function of s and the vector $\vec{\mu}$, which may be considered as a vector belonging to the hyperplane Φ_s normal to ℓ at the point s. If we are given $(s, y_1, \ldots y_m)$, then we are also given the vectors $\vec{e}_1, \ldots \vec{e}_m$, which form a basis in each Φ_s, $s \in \ell$. The components $\vec{e}_1, \ldots \vec{e}_m$ of the vectors $\vec{\mu} \in \Phi_s\,(\vec{\mu} = \sum_{s=1}^{m} \mu_s \vec{\ell}_s)$ may be considered

*This geometrical object is a fiber space with m-dimensional vector fiber (cf. [9]).

coordinates in Φ_s, induced by the coordinate system $s, y_1, \ldots y_m$. Equation (3.13) can be written in the form

$$2i\frac{\mathcal{D}\mathcal{U}}{ds}+\Delta\mathcal{U}-(K(s)\vec{\mu},\vec{\mu})\mathcal{U}=0, \tag{3.14}$$

where $\frac{\mathcal{D}}{ds}$ is the "covariant" derivative of the function $\mathcal{U}$, which is understood as the limit

$$\frac{\mathcal{D}\mathcal{U}(s,\vec{\mu})}{ds}=\lim_{\Delta s\to 0}\frac{\mathcal{U}(s+\Delta s,\vec{\mu}_{\Delta s})-\mathcal{U}(s,\vec{\mu})}{\Delta s}. \tag{3.15}$$

Here $\vec{\mu}_{\Delta s}$ is the result of translating the vector parallel along ℓ from the point $s\in\ell$ to the point $s+\Delta s\in\ell$, Δ is the Laplace operator in the Euclidean hyperplane Φ_s, and $(K(s)\vec{\mu},\vec{\mu})$ is a quadratic form on Φ_s in the coordinate system $s, \mu_1, \ldots \mu_m$, which has the form

$$\sum_{j,h=1}^{m} K_{jh}\mu_j\mu_h .$$

Our problem is to find solutions of equation (3.14) in the normal bundle Ξ which satisfy the following conditions:

$$\max_{|\vec{\mu}|=R}|\mathcal{U}(s,\vec{\mu})|\xrightarrow[R\to\infty]{}0;\quad \mathcal{U}\not\equiv 0, \tag{3.16}$$

$$\mathcal{U}(s+\mathcal{L},\vec{\mu})=e^{i\varkappa}\,\mathcal{U}(s,\vec{\mu}). \tag{3.17}$$

In analogy with the theory of ordinary linear differential equations, we shall call solutions of problem (3.14), (3.16), (3.17) Floquet solutions, and we shall call $\varkappa$ the Floquet exponent. It will be seen in the following that this problem can be solved in a certain sense. The possibility of such a solution requires that the geodesic ℓ satisfy the condition of "stability in the first approximation" (a concept introduced by V. S. Buldyrev [5]). In this case it turns out that there exists a countable set of Floquet exponents to which there correspond solutions of problem (3.14), (3.16), (3.17). To each such exponent there corresponds a finite set of Floquet solutions.

If a solution of problem (3.14), (3.16), (3.17) has been found, then it is not hard to obtain the asymptotic behavior of the eigenvalues from $\varkappa$ by requiring that the eigenfunction [cf. formula (3.7)] be periodic in s; this ensures the uniqueness of $\mathcal{U}$.

Formulas (3.7) and (3.17), the uniqueness condition, give the following expression for the eigenvalues $K = K_p$:

$$K_p \approx \frac{1}{\mathcal{L}}(2\pi p-\varkappa) \tag{3.18}$$

(p is here an integer $\gg 1$).

§4. The Jacobi Equation for the Geodesic ℓ

The geodesic lines play the role of rays in the problem under consideration (cf. §§2-3). In the following a fundamental role will be played by the equations for the rays (i.e., the geodesics) near the geodesic ℓ, that is, the Euler equations for the local ray method.

Let $\{y_j = y_j(s)\}$ be the equation of a geodesic near ℓ. In the basic variational equation for geodesics

$$\delta\int\sqrt{\sum_{i,j=0}^{m} g_{ij}\frac{dy^i}{ds}\frac{dy^j}{ds}}=0,\quad (y^0,y^1,\ldots y^m)=(s,y_1,\ldots y_m) \tag{4.1}$$

we shall assume that y_j and $\frac{dy_i}{ds}$ are small of first order. Using relations (3.1), (3.4), and (3.5) and neglecting terms which are small of second order and higher, we obtain

$$\int\sqrt{\sum_{i,j=0}^{m} g_{ij}\frac{dy^i}{ds}\frac{dy^j}{ds}} \approx \int\sqrt{1-\sum_{i,j=1}^{m} K_{ij}(s)y^i y^j+\dots} \approx \int\Big(1+\frac{1}{2}\sum_{j=1}^{m}\Big(\frac{dy_j}{ds}\Big)^2-\frac{1}{2}\sum_{i,j=1}^{m} K_{ij}(s)y_i y_j\Big)ds. \tag{4.2}$$

The Euler equations for the functional (4.2) have the form

$$\frac{d^2 y_j}{ds^2}+\sum_{i=1}^{m} K_{ij}(s)y_i=0. \tag{4.3}$$

The system of equations (4.3) can be written in a convenient invariant form. Let $\vec{e}_1,\dots,\vec{e}_m$ be the vectors introduced in §3. We put

$$\vec{J}=\sum_{j=1}^{m} y_j(s)\vec{e}_j. \tag{4.4}$$

The vector $\vec{J}$ satisfies the equation

$$\frac{\mathcal{D}^2\vec{J}}{ds^2}+K(s)\vec{J}=0.^* \tag{4.5}$$

Here $\frac{\mathcal{D}}{ds}$ denotes covariant differentiation:

$$\frac{\mathcal{D}\vec{J}}{ds}=\lim_{\Delta s\to 0}\frac{\vec{J}_{\Delta s}(s+\Delta s)-\vec{J}(s)}{\Delta s}. \tag{4.6}$$

It is assumed that the vector $\vec{J}$ is defined for all $s\in\ell$; the vector $\vec{J}_{\Delta s}(s+\Delta s)$ is a vector belonging to the hyperplane Φ_s (normal to ℓ at the point s) which is obtained by translating the vector $\vec{J}(s+\Delta s)$ parallel along ℓ from the point $s+\Delta s$ to the point s; $K(s)$ is a linear operator from Φ_s to Φ_s which in the basis $\vec{e}_1,\dots\vec{e}_m$ has the form

$$K\vec{J}=K\sum_{j=1}^{m} y_j(s)\vec{e}_j=\sum_{j,\tau=1}^{m} K_{\tau j}(s)y_j\vec{e}_\tau \tag{4.7}$$

Since the matrix $\|K_{\tau j}\|$ is symmetric and real [cf. formula (3.5)], the operator K is self-adjoint:

$$(K\vec{J},\vec{V})=(\vec{J},K\vec{V}),\quad \vec{J},\vec{V}\in F_s. \tag{4.8}$$

The scalar product of two vectors $\vec{J}=\sum_{j=1}^{m} J_j\vec{e}_j$ and $\vec{V}=\sum_{j=1}^{m} V_j\vec{e}_j$ is given by

$$(\vec{J},\vec{V})=\sum_{\tau=1}^{m} J_\tau\bar{V}_\tau. \tag{4.9}$$

The bar indicates complex conjugation (we must consider not only real, but also complex solutions of the vector equation (4.5)).

The properties of the vector equation (4.5) are in many respects analogous to those of the scalar equation of second order $y''_{xx}+p(x)y=0$. For example, the classical result of the independence of x of the Wronskian of two solutions of this scalar equation has the following two analogues:

1) the Wronskian of any $2m$ solutions of equation (4.5) is constant;

2) if $\vec{J}$ and $\vec{V}$ are two solutions of equation (4.5), then

$$\Big(\frac{\mathcal{D}\vec{J}}{ds},\vec{V}\Big)-\Big(\vec{J},\frac{\mathcal{D}\vec{V}}{ds}\Big)=\mathrm{const}. \tag{4.10}$$

*The Jacobi equation is written in this form in the monograph [10].

To prove (4.10), we differentiate the left side with respect to s and use: 1) the equation

$$\frac{d}{ds}(\vec{\mathcal{V}}_1, \vec{\mathcal{V}}_2) = \left(\frac{\mathcal{D}\vec{\mathcal{V}}_1}{ds}, \vec{\mathcal{V}}_2\right) + \left(\vec{\mathcal{V}}_1, \frac{\mathcal{D}\vec{\mathcal{V}}_2}{ds}\right),$$

which is true for any pair of vectors continuously differentiable with respect to s, 2) equation (4.5), and 3) formula (4.8).

Equation (4.5) for the ray "in first approximation" $\vec{J} = \vec{J}(s)$ is the well-known Jacobi equation for the geodesic. [The Jacobi equation is defined as the Euler equation for the second variation of the functional, and the expression (4.2), for which (4.5) is the Euler equation, differs from the second variation of the functional $\int d\sigma = \int \sqrt{g_{ij} dy^i dy^j}$ only by a constant term.]

We call the geodesic ℓ stable if equation (4.5) has no unbounded solutions for $s \to \pm\infty$.

All our subsequent considerations are meaningful only if ℓ is stable. We note that the definition of stability of ℓ is in complete agreement with the concept of ray stability in the first approximation, which plays a fundamental role in [5].

We shall now derive an important criterion for the stability of ℓ.

The set of solutions of equation (4.5) forms a $2m$ -dimensional linear space (since $\vec{J}$ is an m-dimensional vector, and equation (4.5) is of second order). Since the geodesic ℓ is closed, if s is moved a distance $\mathcal{L}$ ($\mathcal{L}$ is the length of ℓ), we obtain a certain one-to-one mapping of this space into itself. The operator S which performs this mapping is called the monodromy operator. The eigenvectors $\vec{J}$, that is, the solutions of equation (4.5) which satisfy the conditions

$$S\vec{J}(s) \equiv \vec{J}(s+\mathcal{L}) = \lambda \vec{J}(s) , \tag{4.11}$$

we call Floquet solutions. The following assertion is almost obvious.

1°. In order that ℓ be a stable geodesic it is necessary that for any Floquet solution the constant λ be equal to 1 in modulus.

Indeed, if this is not the case, then for $s \to +\infty$ (if $|\lambda| > 1$) or for $s \to -\infty$ (if $|\lambda| < 1$), the Floquet solution, which satisfies condition (4.11), will be unbounded. We shall now indicate how λ is found.

Let $\vec{J}_1, \dots \vec{J}_{2m}$ be a fundamental system of solutions of equation (4.5). $\vec{J}_1, \dots \vec{J}_{2m}$ may be assumed to form a basis in the space of solutions of this equation (i.e., any solution is a linear combination of the $\vec{J}_j$). Let the operator S be represented as the matrix $S = \| S_{j\tau} \|$ with respect to this basis (that is, under displacement a distance $\mathcal{L}$ the solution $\vec{J}_j(s)$ goes over into the solution

$$\vec{J}_j(s+\mathcal{L}) = S\vec{J}_j = \sum_{\tau=1}^{2m} \vec{J}_\tau S_{j\tau} \quad (j = 1, 2, \dots, m). \tag{4.12}$$

If $\vec{J} = A_j \vec{J}_j(s)$, ($A_j = \text{const}$) is a Floquet solution, then clearly λ is an eigenvalue of the matrix $\| S_{j\tau} \|$; that is, λ is a solution of an equation of degree $2m$:

$$P(\lambda) = \det(\| S_{j\tau} \| - \lambda \cdot I) \tag{4.13}$$

(I is the identity matrix). Thus, condition 1° is equivalent to the following condition.

1. In order that ℓ be stable it is necessary that all roots of equation (4.13) be equal to one in modulus. It is not difficult to show that the polynomial $P(\lambda)$ does not depend on the choice of the fundamental system of solutions: in replacing one fundamental system of solutions by another we replace $\| S_{j\tau} \|$ by a matrix similar to it. For a real fundamental system of solutions [such solutions exist, since equation (4.5) is real] the matrix $S_{j\tau}$, and hence the coefficients of the polynomial $P(\lambda)$ will be real.

In a manner similar to what is done in the case of a scalar equation of Hille type (see [11]), in the case of equation (4.5) it is not difficult to show that

$$\det S = \det \| S_{j\tau} \| = 1. \tag{4.14}$$

However, in our case there is a more general and important theorem. The polynomial $P(\lambda)$ is recurrent, that is, the coefficients of $\lambda^{2n-\tau}$ and λ^{τ} are the same, $\tau = 0, 1, 2, \dots 2m$, or what is the same thing

$$P(\lambda) = \lambda^{2m} P\left(\frac{1}{\lambda}\right). \tag{4.15}$$

To prove the theorem we need a relatively simple lemma: the matrix

$$M = \left\| \left(\frac{\mathcal{D}\vec{\jmath}_j}{ds}, \vec{\jmath}_\tau\right) - \left(\vec{\jmath}_j, \frac{\mathcal{D}\vec{\jmath}_\tau}{ds}\right) \right\| \tag{4.16}$$

is nonsingular if the vectors $\vec{\jmath}_1, \dots \vec{\jmath}_{2m}$ form a fundamental system of solutions.

We proceed to the proof of the lemma. We replace the fundamental system of solutions $\vec{\jmath}_1, \dots \vec{\jmath}_{2m}$ by another fundamental system of solutions $\vec{\jmath}'_1, \dots \vec{\jmath}'_{2m}$, where the vectors $\vec{\jmath}'_j$ are linear combinations of the vectors $\vec{\jmath}_j$. This can be written in matrix form: let the columns of the $m \times 2m$ matrices $\jmath$ and $\jmath'$ consist of the vectors $\vec{\jmath}_j$ and $\vec{\jmath}'_j$ respectively

$$\jmath = \| \vec{\jmath}_1, \dots \vec{\jmath}_{2m} \|, \quad \jmath' = \| \vec{\jmath}'_1, \dots \vec{\jmath}'_{2m} \| \ ; \tag{4.17}$$

then there exists a $2m \times 2m$ matrix $C (= \mathrm{const})$ such that

$$\det C \neq 0, \quad \jmath' = \jmath C. \tag{4.18}$$

The matrix M can obviously be written

$$M = \jmath^* \frac{\mathcal{D}\jmath}{ds} - \left(\frac{\mathcal{D}\jmath}{ds}\right)^* \jmath, \tag{4.19}$$

where $\jmath^*$ (or $\left(\frac{\mathcal{D}\jmath}{ds}\right)^*$ is obtained from $\jmath$ (or $\frac{\mathcal{D}\jmath}{ds}$) by interchanging rows and columns and replacing all elements by their complex conjugates.

Relations (4.18) and (4.19) imply that

$$M' = C^* M, \quad \det C \neq 0, \quad M' = \jmath'^* \frac{\mathcal{D}\jmath'}{ds} - \left(\frac{\mathcal{D}\jmath'}{ds}\right)^* \jmath \tag{4.20}$$

(C^* is the Hermitian conjugate of the matrix C).

Formula (4.20) and the fact that the matrix M is constant [cf. formulas (4.10) and (4.16)] imply that if $\det M \neq 0$ for some fundamental system of solutions at some point, then it cannot vanish at any point for any fundamental system of solutions.

To complete the proof of the lemma it is sufficient to note that $\det M \neq 0$ for the fundamental system of solutions $\vec{\jmath}_j$, which satisfy the initial conditions

$$\vec{\jmath}_j \Big|_{s=0} = \begin{cases} \vec{e}_j, & j = 1, 2, \dots m, \\ 0, & j = m+1, \dots 2m, \end{cases} \quad \frac{\mathcal{D}\vec{\jmath}_j}{ds} = \begin{cases} 0, & j = 1, 2, \dots m, \\ \vec{e}_{j-m}, & j = m+1, \dots 2m \end{cases}$$

(the $\vec{e}_j$ are the vectors introduced in §3).

We now proceed to the proof of the theorem. Equation (4.12) can be written in matrix form:

$$\mathcal{J}(s+\mathcal{L}) = \mathcal{J}(s)\,S; \quad \mathcal{J} = \| \vec{\mathcal{J}}_1, \ldots \vec{\mathcal{J}}_{2m} \|, \qquad (4.21)$$

$$S = \| S_{jk} \|.$$

From formula (4.19) and the constancy of the matrix M it follows that

$$S^* M S = M, \qquad (4.22)$$

whence

$$M S M^{-1} = S^{*-1}$$

(here we have used the fact that $\det M \neq 0$). Subtracting $\lambda \cdot I$ from both sides of this equation (λ is an arbitrary number, and I is the identity matrix), we obtain

$$M(S - \lambda \cdot I) M^{-1} = S^{*-1} - \lambda \cdot I = (-\lambda)(S^* - \tfrac{1}{\lambda} I) \cdot S^{*-1},$$

whence, on going over to the determinants:

$$P(\lambda) = \det(S - \lambda \cdot I) = (-\lambda)^{2m} \det(S^* - \tfrac{1}{\lambda}) \det S^{*-1} = \lambda^{2m} P(\tfrac{1}{\lambda}). \qquad (4.23)$$

We now make use of the fact that the coefficients of the polynomial $P(\lambda)$ are real and also the equation

$$\det S^{*-1} = \frac{1}{\det S^*} = \frac{1}{\det S} = 1.$$

[cf. formula (4.14)].

Equations (4.23) and (4.15) are identical, and this completes the proof of our theorem.

Thus, in order to verify that condition 1° or condition 1 is satisfied, it is not necessary to solve the equation $P(\lambda) = 0$, of degree $2m$, but rather the equation of degree m which is obtained by introducing the new unknown $\lambda_1 = \frac{1}{2}(\lambda + \frac{1}{\lambda})$. It is well known that if the polynomial $P(\lambda)$ is recurrent then the expression $\lambda^{-m} P(\lambda)$ is a polynomial of degree m in $\lambda_1 = \frac{1}{2}(\lambda + \frac{1}{\lambda})$ with real coefficients. Verification of condition 1 reduces to checking that the roots of this polynomial lie in the interval $-1 \leqslant \lambda_1 \leqslant 1$.

Suppose first of all that all roots of the polynomial $P(\lambda)$ are simple. This is equivalent to the assumption that all roots $\lambda_1 = \frac{1}{2}(\lambda + \frac{1}{\lambda})$ of the polynomial $P(\lambda)$ are distinct and lie in the open interval $-1 < \lambda < +1$ (condition 1° or 1 is assumed to be satisfied). The presence of a root $\lambda_1 = \pm 1$ now leads to a root $\lambda = \pm 1$, having multiplicity two; therefore the endpoints of the interval $-1 \leqslant \lambda_1 \leqslant 1$ must be excluded.

In the absence of multiple roots of the polynomial $P(\lambda)$ condition 1° (or 1) is sufficient for the stability of the geodesic. Indeed, in the case under consideration the Floquet solutions form a complete system of $2m$ linearly independent solutions which are bounded on the entire axis (since all the λ are equal to 1 in modulus), and any solution of equation (4.5) is a linear combination of these with constant coefficients.

We shall now normalize these Floquet solutions in a manner convenient for subsequent considerations. We note first of all that for any two Floquet solutions $\vec{\mathcal{J}}$ and $\vec{V}$, to which there correspond distinct constants $e^{\alpha i}$ and $e^{\beta i}$ ($e^{\alpha i} \neq e^{\beta i}$),

$$\Pi(\vec{\mathcal{J}}, \vec{V}) \equiv (\frac{\mathcal{D}\vec{\mathcal{J}}}{ds}, \vec{V}) - (\vec{\mathcal{J}}, \frac{\mathcal{D}\vec{V}}{ds}) = 0. \qquad (4.24)$$

Indeed, by equation (4.10) Π does not depend on s; on the other hand, by (4.11) and the definition of $\Pi(s)$

$$\Pi|_s = \Pi|_{s+\mathcal{L}} = e^{+\alpha i} \cdot e^{\beta i} \Pi|_s$$

whence $\Pi \equiv 0$.

If now $\vec{\jmath} = \vec{v} \neq 0$ is a Floquet solution, then in the case being considered it can be shown that $\Pi(\vec{\jmath}, \vec{\jmath}) \neq 0$. Indeed, forming the matrix M (cf. formula (4.16)) and taking Floquet solutions as $\vec{\jmath}$, it is easy to see by using property (4.24) of the function $\Pi(s)$, that $M(s)$ is a diagonal matrix.

If now for some Floquet solution $\vec{\jmath}_j$ the function $\Pi(s)$ [cf. (4.24)] with $\vec{\jmath} = \vec{v} = \vec{\jmath}_j$ were zero, then the determinant of the matrix $M(s)$ would be zero, which is impossible by the preceding lemma and the fact that the Floquet solutions form a fundamental system of solutions.

Suppose now that $\vec{\jmath}_i \neq 0$ is some Floquet solution ($j = 1, 2, \dots m$). There are two possibilities: either $\frac{1}{i}\Pi(\vec{\jmath}_j, \vec{\jmath}_j) > 0$ or $\frac{1}{i}\Pi(\vec{\jmath}_j, \vec{\jmath}_j) < 0$ [the fact that $\Pi(\vec{\jmath}_j, \vec{\jmath}_j)$ is pure imaginary follows from (4.24)], but $\Pi(\vec{\jmath}_j, \vec{\jmath}_j) \neq 0$.

In the first (second) case we put

$$\vec{z}_j = \vec{\jmath}_j \left[\tfrac{1}{i}\Pi(\vec{\jmath}_j, \vec{\jmath}_j)\right]^{-\frac{1}{2}}, \qquad ([\dots]^{-\frac{1}{2}} > 0)$$

$$\vec{z}_j = \vec{\jmath}_j^{\,*} \left[\tfrac{1}{i}\Pi(\vec{\jmath}_j^{\,*}, \vec{\jmath}_j^{\,*})\right]^{-\frac{1}{2}}, \qquad ([\dots]^{-\frac{1}{2}} > 0)$$

(here $\vec{\jmath}_j^{\,*}$ is a vector with components which are complex conjugates of the components of $\vec{\jmath}_j$).

The vectors $\vec{z}_1, \dots \vec{z}_m, \vec{z}_1^{\,*}, \dots \vec{z}_m^{\,*}$ form a fundamental system of solutions, since these are Floquet solutions corresponding to the $2m$ distinct roots of the polynomial $P(\lambda) = 0$.

The vectors z_j satisfy the relations

$$\left(\frac{\mathcal{D}\vec{z}_j}{ds}, \vec{z}_h\right) - \left(\vec{z}_j, \frac{\mathcal{D}\vec{z}_h}{ds}\right) = i\,\delta_{jh} \tag{4.25}$$

(δ_{jh} is the Kronecker symbol)

$$\left(\frac{\mathcal{D}\vec{z}_j}{ds}, \vec{z}_h^{\,*}\right) - \left(\vec{z}_j, \frac{\mathcal{D}\vec{z}_h^{\,*}}{ds}\right) = 0. \tag{4.26}$$

Introducing an $m \times m$, matrix z with the vectors $\vec{z}_j$ $j = 1, 2, \dots m$, as columns, conditions (4.25) and (4.26) can be written in the form of two matrix equations:

$$z^* \frac{\mathcal{D}z}{ds} - \left(\frac{\mathcal{D}z}{ds}\right)^* z = i, \tag{4.27}$$

$$z^T \frac{\mathcal{D}z}{ds} - \left(\frac{\mathcal{D}z}{ds}\right)^T z = 0, \qquad z = \|\vec{z}_1, \vec{z}_2, \dots \vec{z}_m\| \tag{4.28}$$

(the asterisk denotes the Hermitian conjugate; the symbol T denotes the transpose).

If not all the roots of the polynomial $P(\lambda)$ are simple, then the condition 1 (or 1°) is not sufficient for the stability of the geodesic ℓ. In this case we have the following condition.

2. A necessary and sufficient condition for the stability of ℓ is condition 1 and the absence of associated eigenvectors of the monodromy operator S.

We shall prove the necessity of condition 2.

Let $\vec{\jmath}$ be an eigenvector of the monodromy operator S, and let $\vec{\jmath}_1$ be an associated vector, that is,

$\vec{J}$ and $\vec{J}_1$ are solutions of equation (4.5) such that

$$\vec{J}(s+\mathcal{L}) = \lambda \vec{J}(s),$$
$$\vec{J}_1(s+\mathcal{L}) = \lambda \vec{J}_1(s) + \vec{J}(s), \quad \vec{J}(s) \not\equiv 0. \tag{4.29}$$

Using formula (4.29) M times, it is easy to show that

$$\vec{J}_1(s+M\mathcal{L}) = \lambda^M \vec{J}_1(s) + M\lambda^{M-1}\vec{J}(s). \tag{4.30}$$

Since M can be arbitrarily large and $\vec{J}(s) \not\equiv 0$, we conclude from (4.30) that the vector $\vec{J}_1(s)$ is unbounded on the semiaxis $0 \leqslant s < +\infty$.

The sufficiency of condition 2 follows easily from the fact that if it is satisfied the Floquet solutions form a fundamental system of solutions which are bounded for $-\infty < s < +\infty$.

We note that condition 2 may be reformulated as follows: all the roots of the polynomial $P(\lambda) = \det(S-\lambda 1)$ must be equal to one in modulus, and the elementary divisors must be simple.

Suppose that not all roots of the polynomial $P(\lambda)$ are simple and that condition 2 is satisfied. Then it is possible (as in the case of simple roots of $P(\lambda)$) to construct a matrix z, with columns which are Floquet solutions $\vec{z}_j$, such that equations (4.25)-(4.28) are satisfied. It is interesting to note that the choice of constants $\lambda_j = \exp i\alpha_j$ corresponding to the Floquet solutions $\vec{z}_j$ will be uniquely determined in the case (as in the case of simple roots of $P(\lambda)$) but the vectors $\vec{z}_j$ themselves will not be uniquely determined.

We now proceed to the construction of the vectors $\vec{z}_j$. Let $\vec{J}_1, \dots \vec{J}_{2m}$ be vectors which are Floquet solutions of equation (4.5) and which form a fundamental system. We form the $(m \times 2m)$ matrix J with columns consisting of the vectors $\vec{J}_j$.

We shall assume that solutions corresponding to the same Floquet exponent λ_0 occupy neighboring columns (the number of such columns is equal to the multiplicity of the root λ_0 of the polynomial $P(\lambda)$). It follows easily from (4.19) that the matrix M is quasidiagonal: $M = \left\| \begin{smallmatrix} \boxplus & 0 \\ 0 & \boxplus \end{smallmatrix} \right\|$. A single "box" contains solutions J_j and J_r for which $\lambda_j = \lambda_r$.

Replacing, if necessary, Floquet solutions corresponding to the same λ by linear combinations of such solutions, the matrix M can be brought to diagonal form.

Let $h_m > 1$ be the multiplicity of the root $\lambda = e^{i\alpha} \neq \pm 1$ of the polynomial $\dot{P}(\lambda)$. By condition 2 there exist precisely h linearly independent Floquet solutions $\vec{v}_1, \dots \vec{v}_h$ such that $\vec{v}_j(s+\mathcal{L}) = \lambda_j \vec{v}(s)$. We shall assume that the vectors $\vec{v}_1, \dots \vec{v}_h$ are contained among the $\vec{J}_1, \dots \vec{J}_{2m}$. To the vectors $\vec{v}_j$ there corresponds a square "box" on the diagonal of the matrix M:

$$H = \left\| \left(\frac{\mathcal{D}\vec{v}_j}{ds}, \vec{v}_r\right) - \left(\vec{v}_j, \frac{\mathcal{D}\vec{v}_r}{ds}\right) \right\|, \quad j, r = 1, 2, \dots h.$$

If the vectors $\vec{v}_j$ are replaced by a linear combination of them $\vec{v}_j'$,

$$\vec{v}_j' = \sum_{r=1}^{n} C_{rj} v_r,$$

then the matrix H is replaced by the matrix C^*HC $(C = \|C_{rj}\|)$. Since $\frac{1}{i}H$ is a Hermitian matrix, the constant matrix C (or, what is the same thing, the Floquet solutions $\vec{v}_j'$) can be chosen such that the matrix $\frac{1}{i}C^*HC$ is diagonal, and on the diagonal there will be τ_1 ones and τ_2 minus ones with $\tau_1 + \tau_2 = h$. For $h > 1$ there exist uncountably many matrices C effecting the diagonalization of H; however, because of the law of inertia for quadratic forms the numbers τ_1 and τ_2 with $\tau_1 + \tau_2 = h$ are unique.

In a similar manner the analogue H_1 of H, corresponding to Floquet solutions for the root $\bar{\lambda}=e^{-\alpha i}$ of the polynomial $P(\lambda)$ can be diagonalized. In particular, if in this case we take the solutions $\vec{v}_j^{*1}$ as the Floquet solutions (here $\vec{v}_j'(j=1,2,\dots h)$ are the solutions diagonalizing the matrix H, and the asterisk indicates that all components of the corresponding vector have been replaced by their complex conjugates), then we find that the diagonalized matrix $\frac{1}{i}H_1$ has $\varkappa_2$ ones and $\varkappa_1$ minus ones along the diagonal, since

$$\left(\frac{\mathcal{D}\vec{v}_j^{*1}}{ds},\vec{v}_\tau^{*1}\right)-\left(\vec{v}_j^{*1},\frac{\mathcal{D}\vec{v}_\tau^{*1}}{ds}\right)=\overline{\left(\frac{\mathcal{D}\vec{v}_j'}{ds},\vec{v}_\tau'\right)-\left(\vec{v}_j',\frac{\mathcal{D}v_\tau'}{ds}\right)}=-\left[\left(\frac{\mathcal{D}\vec{v}_j'}{ds},\vec{v}_\tau'\right)-\left(\vec{v}_j',\frac{\mathcal{D}\vec{v}_\tau'}{ds}\right)\right]$$

(the bar indicates complex conjugation).

If $\lambda=+1$ or -1, then the corresponding matrix $\frac{1}{i}H$ can also be diagonalized. Let $\varkappa_1$ be the number of ones on the diagonal and $\varkappa_2$ the number of minus ones. Replacing $\vec{v}_j'$ by $\vec{v}_j^{*1}$ we again obtain a diagonalization of H in which the roles of the numbers $\varkappa_1$ and $\varkappa_2$ are interchanged. This implies that $\varkappa_1=\varkappa_2$, because of the law of inertia of quadratic forms. We carry out a similar diagonalization in each "box" of the matrix M.

We shall now choose a fundamental system of solutions in the following manner: for the first m vectors we take those vectors $\vec{z}_1,\dots\vec{z}_m$ (of the fundamental system of solutions obtained during the process of diagonalizing the matrix M) which satisfy the condition

$$\left(\frac{\mathcal{D}\vec{z}_j}{ds},\vec{z}_j\right)-\left(\vec{z}_j,\frac{\mathcal{D}\vec{z}_j}{ds}\right)=i,\qquad j=1,2,\dots m. \tag{4.31}$$

Since M is diagonal, they will also satisfy the conditions

$$\left(\frac{\mathcal{D}\vec{z}_j}{ds},\vec{z}_\tau\right)-\left(\vec{z}_j,\frac{\mathcal{D}\vec{z}_\tau}{ds}\right)=0,\qquad j\neq\tau. \tag{4.32}$$

For the remaining vectors of the fundamental system of solutions, we take $\vec{z}_1^*,\vec{z}_2^*,\dots\vec{z}_m^*$. Clearly

$$\left(\frac{\mathcal{D}\vec{z}_j}{ds},\vec{z}_\tau^*\right)-\left(\vec{z}_j,\frac{\mathcal{D}\vec{z}_\tau^*}{ds}\right)=0,\qquad j,\tau=1,2,\dots m. \tag{4.33}$$

The relations (4.31) and (4.33) coincide with (4.25) and (4.26). This completes the construction of the required vectors $\vec{z}_j$.

§5. Solutions of the Problem (3.14), (3.16), (3.17)

We seek solutions of the parabolic equation (3.14) in the form

$$u=S(s)e^{\frac{i}{2}(\Gamma\vec{\mu},\vec{\mu})}, \tag{5.1}$$

where $S(s)$ is a function depending only on s and $(\Gamma\vec{\mu},\vec{\mu})$ is a quadratic form on the m-dimensional hyperplane Φ_s, normal to ℓ at the point s.

In the coordinate system $s,\mu_1,\dots\mu_m$ ($\mu_1,\dots\mu_m$ are the components of the vector $\vec{\mu}$ on the hyperplane Φ_s) the quadratic form $(\Gamma\vec{\mu},\vec{\mu})$ can be written in the form

$$\sum_{i,j=1}^{m}\Gamma_{ij}(s)\mu_i\mu_j,\quad\text{where } \Gamma=\|\Gamma_{ij}(s)\|,\quad i,j=1,2,\dots m \tag{5.2}$$

is a symmetric matrix. Substituting (5.1) in the parabolic equation (3.11), we obtain for Γ a matrix Riccati equation:

$$\Gamma'+\Gamma^2+K=0. \tag{5.3}$$

Here $K = \|K_{ij}(s)\| = \|R_{0i0j}(s)\|$ is an $m \times m$ matrix [cf. formula (3.5)].

For $S(s)$ we obtain the equation

$$2S' + S\,sp\,\Gamma = 0 \tag{5.4}$$

($Sp\,\Gamma$, as usual, denotes the sum of the diagonal elements of the matrix Γ).

The matrix Riccati equation reduces to a system of ordinary linear equations (in the same manner as its scalar analogue).

We shall seek a solution of equation (5.3) in the form

$$\Gamma = z' \cdot z^{-1} = \frac{dz}{ds} z^{-1}, \tag{5.5}$$

where z is a new unknown matrix. Substituting (5.5) in (5.3) and using the obvious identity

$$z \frac{dz^{-1}}{ds} + \frac{dz}{ds} z^{-1} = \frac{d}{ds} z z^{-1} \equiv 0,$$

we find for the matrix z the equation

$$\frac{d^2 z}{ds^2} + Kz = 0. \tag{5.6}$$

Equation (5.6) means that the following equations are satisfied:

$$\frac{d^2 z_{hj}}{ds^2} + \sum_{j=1}^{m} K_{hj}(s) z_{jh} = 0, \quad h, j = 1, \dots m. \tag{5.7}$$

Equation (5.7) for fixed j and $h = 1, 2, \dots m$ coincides with equation (4.5) written in coordinate form if we put $\vec{J} = (J_1, \dots J_m) = (z_{1j}, z_{2j}, \dots z_{mj}) = \vec{z}_j$. We thus obtain a solution of equation (5.3) if we take for the matrix Γ the expression (5.5), with z a matrix whose columns are solutions of equation (4.5). For this matrix z we take the matrix $\|\vec{z}_1, \dots \vec{z}_m\|$, with columns which are Floquet solutions $\vec{z}_j$ of equation (4.5) (cf. formulas (4.25)-(4.28)).

We now turn to equation (5.4). Inserting the expression for Γ, we find easily that, up to a constant multiple, S is equal to the following:

$$S(s) = exp\left[-\frac{1}{2} \int_0^s sp\, z' z^{-1} ds\right].$$

Using the well-known formula for the derivative of a determinant (cf. [12]),

$$\frac{d}{ds} \det z = sp(z' z^{-1}) \det z,$$

we obtain (again up to a multiplicative constant)

$$S(s) = \frac{1}{\sqrt{\det z}}. \tag{5.8}$$

Formulas (5.1), (5.5), and (5.8) now give

$$u_0 = \frac{1}{\sqrt{\det z}} e^{\frac{i}{2}(z' z^{-1} \mu, \mu)} \; * \tag{5.9}$$

*This formula is a good illustration of the well-known fact that the semiclassical approximation for the Schrödinger equation with a quadratic potential is an exact solution. Concerning the integration of the Schrödinger equation see [13] and [14].

It must now be shown that $\mathcal{U}_0$, is actually one of the solutions of problem (3.14), (3.16), (3.17).

We shall show first of all that formula (5.5) defines a symmetric matrix. Let us return to relation (4.28). In the coordinates $(s, \mu_1, \ldots \mu_m)$ it is written

$$Z^T Z' - Z'^T Z = 0.$$

Multiplying this equation on the left by $(Z^T)^{-1}$ and on the right by Z^{-1}, we obtain

$$Z' Z^{-1} - (Z^T)^{-1} (Z')^T$$

or

$$\Gamma = Z' Z^{-1} = (Z' Z^{-1})^T = \Gamma^T,$$

i.e., the matrix of (5.5) is indeed symmetric.

To prove property (3.16) it is sufficient to show that

$$Im\ \Gamma = \frac{1}{2i}(\Gamma - \Gamma^*) \tag{5.10}$$

is a positive definite matrix.

We use formula (4.27)

$$Z^* Z' - (Z')^* Z = i \cdot 1.$$

Multiplying this equation on the left by $(Z^*)^{-1}$ and on the right by Z^{-1}, we obtain

$$Z' Z^{-1} - Z^{*-1} Z'^* = i Z^{*-1} Z^{-1}$$

or

$$\Gamma - \Gamma^* \equiv 2i\, Im\ \Gamma = i (Z Z^*)^{-1}, \tag{5.11}$$

whence it follows that the matrix (5.10) is positive definite.

It remains to show that the function (5.9) satisfies condition (3.17).

Displacing the vectors $\vec{Z}_j$ a distance $\mathcal{L}$ ($\mathcal{L}$ is the length of ℓ), they acquire multipliers $exp\ (i\alpha_j)$; therefore,

$$\begin{aligned} Z(s+\mathcal{L}) &= Z(s) I(\alpha), \\ Z'(s+\mathcal{L}) &= Z'(s) I(\alpha). \end{aligned} \tag{5.12}$$

where

$$I(\alpha) = \left\| \begin{matrix} e^{i\alpha_1} & & & 0 \\ & e^{i\alpha_2} & & \\ & & \ddots & \\ 0 & & & e^{i\alpha_m} \end{matrix} \right\|. \tag{5.13}$$

Equations (5.12) and (5.13) imply that

$$\Gamma(s+\mathcal{L}) \equiv \Gamma(s) \tag{5.14}$$

and

$$\mathcal{U}_0(s+\mathcal{L}) = e^{i\varkappa_0}\, \mathcal{U}_0(s). \tag{5.15}$$

where

$$\varkappa_0 = -\frac{1}{2}(\alpha_1 + \alpha_2 + \ldots + \alpha_m), \tag{5.16}$$

that is, $\mathcal{U}_0$ is indeed a Floquet solution of the parabolic equation (3.14). We note that, in general, the Floquet exponents are determined up to an integral multiple of 2π. In order that formula (5.16) hold, they must satisfy the condition

$$e^{-\frac{i}{2}(\alpha_1+\ldots+\alpha_m)} = e^{i\varkappa_0} = \frac{S(\mathcal{L})}{S(0)}, \tag{5.17}$$

which may be considered a normalization condition. [We note that the right side of (5.17) is uniquely determined.] We shall assume that the α_i have been chosen such that (5.17) is satisfied.

Starting with the Floquet solution $\mathcal{U}_0$, it is possible to find other Floquet solutions of the parabolic equation (3.14).

We note first of all that any operator of the form

$$\Lambda = \frac{1}{i}\sum_{j=1}^{m} \gamma_j \frac{\partial}{\partial \mu_j} - (\vec{\gamma}', \vec{\mu}) = \frac{1}{i}(\vec{\gamma}, \nabla_\mu) - (\vec{\gamma}', \vec{\mu}), \tag{5.18}$$

where the $\gamma_j^{(s)}$ are components of a vector solving equation (4.5), commutes with the operator $\mathcal{L}$ (cf. formula (3.13)]; this is easy to verify directly.

We now form the following operators:

$$\Lambda_j = \frac{1}{i}(\vec{z}_j, \nabla_\mu) - \left(\frac{\mathcal{D}\vec{z}_j}{ds}, \vec{\mu}\right), \qquad j = 1, \ldots, m \tag{5.19}$$

and

$$\Lambda_j^* = \frac{1}{i}(\vec{z}_j^*, \nabla_\mu) - \left(\frac{\mathcal{D}\vec{z}_j^*}{ds}, \mu\right), \qquad j = 1, \ldots, m. \tag{5.20}$$

They satisfy the commutation relations

$$\left.\begin{array}{l} \Lambda_j \Lambda_h - \Lambda_h \Lambda_j = 0 \\ \Lambda_j^* \Lambda_h^* - \Lambda_h^* \Lambda_j^* = 0 \end{array}\right\}, \tag{5.21}$$

$$\Lambda_j \Lambda_h^* - \Lambda_h^* \Lambda_j = \delta_{hj} \tag{5.22}$$

(δ_{hj} is the Kronecker symbol).

Moreover,

$$\Lambda_j \mathcal{U}_0 = S(s) e^{\frac{i}{2}(\Gamma\mu,\mu)} [(\Gamma z_j, \mu) - (z_j', \mu)] \equiv 0. \tag{5.23}$$

From the fact that $\mathcal{L}$ and Λ_j^* commute it follows that the functions

$$\mathcal{U}_{q_1 \cdots q_m} = (\Lambda_1^*)^{q_1} (\Lambda_2^*)^{q_2} \ldots (\Lambda_m^*)^{q_m} \mathcal{U}_0 = Q_{q_1 \cdots q_m}(\vec{\mu}) \cdot e^{\frac{i}{2}(\Gamma\mu,\mu)} \tag{5.24}$$

will be solutions of the parabolic equation $\mathcal{L}\mathcal{U} = 0$; there the q_i are nonnegative integers, and $Q_{q_1 \cdots q_m}(\vec{\mu})$ is a polynomial in $\mu_1, \ldots \mu_m$ of degree $q_1 + \ldots + q_m$ with coefficients depending on s. (Application of the operators $\Lambda_1^{q_1} \ldots \Lambda_m^{q_m}$ to $\mathcal{U}_0$ produces the trivial solution of the equation $\mathcal{L}\mathcal{U} = 0$).

The function $\mathcal{U}_{q_1,\dots q_m}$ is obviously a Floquet solution of the equation $\mathcal{L}\mathcal{U}=0$, with exponent

$$\varkappa_q=-\left[\frac{1}{2}(\alpha_1+\dots+\alpha_m)+\sum_{j=1}^{m}\alpha_j q_j\right]. \tag{5.25}$$

Indeed, it has already been noted that $\mathcal{U}_{q_1\dots q_m}$ is a solution of the equation $\mathcal{L}\mathcal{U}=0$; the equation

$$\mathcal{U}_{q_1\dots q_m}(s+\mathcal{L},\vec{\mu})=e^{i\delta q}\,\mathcal{U}_{q_1\dots q_m}(s,\vec{\mu})$$

follows from (5.15), the equalities

$$\vec{z}_j^{\,*}(s+\mathcal{L})=e^{i\alpha_j}z_j,\quad \vec{z}_j^{\,*\prime}(s+\mathcal{L})=e^{-i\alpha_j}z_j^{*}(s),$$

and the form of the operators Λ_j^*; the inequalities $\mathcal{U}_{q_1,\dots q_m}\neq 0$ follow from the orthogonality relations below.

It turns out that on any hyperplane Φ_s the functions $\mathcal{U}_{q_1\dots q_m}$ satisfy the following orthogonality relations:

$$\int_{F_s}\mathcal{U}_{q_1,\dots q_m}\overline{\mathcal{U}_{q_1'\dots q_m'}}\,d\vec{\mu}=\iint_{-\infty}^{\infty}\mathcal{U}_{q_1\dots q_m}\overline{\mathcal{U}_{q_1'\dots q_m'}}\,d\mu_1\dots d\mu_m=$$

$$=\begin{cases}0, & (q_1\dots q_m)\neq(q_1'\dots q_m')\\ q_1'q_2'\dots q_m'!\displaystyle\int_{-\infty}^{\infty}\frac{|e^{\frac{i}{2}(\Gamma\mu,\mu)}|^2}{|\det z|}\,d\vec{\mu}=q_1!\,q_2!\dots q_m!\,(4\pi)^{\frac{m}{2}}; & (q_1\dots q_m)=(q_1'\dots q_m').\end{cases} \tag{5.26}$$

The equality (or inequality) $(q_1,\dots q_m)=(q_1',\dots q_m')$ (or $(q_1,\dots q_m)\neq(q_1',\dots q_m')$) means that all the equalities $q_1=q_1',\dots,q_m=q_m'$ hold simultaneously (at least one of these equalities fails to hold).

The equalities (5.26) are easily proved by using integration by parts and formulas (5.21), (5.22), and (5.23).

It follows easily from these formulas that the function $\mathcal{U}_{q_1\dots q_m}$ is an eigenfunction of the elliptic differential operator

$$\sum_{j=1}^{m}\Lambda_j^*\Lambda_j\ , \tag{5.27}$$

(acting on functions defined on Φ_s), corresponding to the eigenvalue

$$q=q_1+\dots+q_m.$$

The linear independence of the polynomials $Q_{q_1\dots q_m}$ [cf. (5.24)] follows from the orthogonality relations (5.26), and since there are as many such polynomials as there are monomials $\mu_1^{q_1}\dots\mu_m^{q_m}$ it follows that any polynomial in the variables $\mu_1\dots\mu_m$, where $(\mu_1,\dots\mu_m)=\vec{\mu}$, is a linear combination of the $Q_{q_1\dots q_m}$.

This implies the completeness of the orthogonal system of functions $\mathcal{U}_{q_1\dots q_m}$. Thus, the functions $\mathcal{U}_{q_1\dots q_m}$ form a complete system of eigenfunctions of the operator (5.27). The completeness of the system of functions $\mathcal{U}_{q_1\dots q_m}$ implies that in a certain sense they form a complete set of Floquet solutions of the equation $\mathcal{L}\mathcal{U}=0$.* Let $\mathcal{U}$ be some Floquet solution of this equation. Multiplying the equation $\mathcal{L}\mathcal{U}=0$ and $\overline{\mathcal{U}}_{q_1\dots q_m}$ and integrating with respect to t_s, we obtain under very general conditions on $\mathcal{U}$:

$$\frac{\partial}{\partial s}\int_{F_s}\mathcal{U}\overline{\mathcal{U}}_{q_1\dots q_m}\,d\vec{\mu}=0. \tag{5.28}$$

Expanding $\mathcal{U}$ in terms of the $\mathcal{U}_{q_1\dots q_m}$ (which is also possible under very general conditions on $\mathcal{U}$), we obtain:

*The completeness proof of the set of solutions we have found is by V. F. Lazutkin.

$$\mathcal{U} = \sum_{p_i \geq 0} C_{p_1 \cdots p_m} \mathcal{U}_{p_1 \cdots p_m}. \tag{5.29}$$

The orthogonality relations (5.26) and formula (5.28) imply that $C_{p_1 \cdots p_m} = const$. The equality

$$\mathcal{U}(s+\mathcal{L}) = e^{i\varkappa} \mathcal{U}(s)$$

and formula (5.29) imply that $\varkappa$ can only have the form (5.25) for some choice of nonnegative integers $q_1 \dots q_m$. Moreover, it can be asserted that

$$\mathcal{U} = const\, \mathcal{U}_{q_1 q_2 \cdots q_m},$$

if the equation

$$\pi\gamma_0 + \sum_{s=1}^{m} \gamma_s \alpha_s = 0 \quad (\gamma_i \text{ - integers}), \tag{5.30}$$

6.

implies that all the $\gamma_s = 0$. If this is not the case then $\mathcal{U}$ is a linear combination (possibly infinite) of the Floquet solutions $\mathcal{U}_{q_1 \cdots q_m}$ corresponding to the same value $e^{i\varkappa_q}$ [$\varkappa_q$ has the form (5.25)].

Formulas (3.18) and (3.7) now give the asymptotic behavior of the eigenvalues and eigenfunctions.

If the $\alpha_1 \dots \alpha_m$ are all distinct and, in addition, the numbers $\pi, \alpha_1, \dots \alpha_m$ are linearly independent over the ring of integers, that is, equation (5.30) implies that all $\gamma_i = 0$, then the set of numbers $q_1 \dots q_m$ uniquely determines an eigenvalue

$$K_{pq_1 \cdots q_m} = 2\pi p + \frac{1}{2}(\alpha_1 + \dots + \alpha_m) + \sum_{j=1}^{m} q_j \alpha_j.$$

In the opposite case [and this will always happen if the polynomial $P(\lambda)$ has multiple roots (cf. §4)], then to different sets $q_1, \dots q_m$ there may correspond identical eigenvalues but different expressions for the eigenfunctions.

In conclusion, we remark that the system of operators satisfying the commutation relations (5.21) and (5.22) plays an important role in quantum field theory. Many of the considerations of the present section [for example, the derivation of formula (5.26)] are carried out in analogy to similar considerations in field theory.

We note further that the functions

$$\mathcal{U}_{q_1 \cdots q_m}(\vec{\mu}) - Q_{q_1 \cdots q_m}(\vec{\mu}) e^{\frac{i}{2}(\Gamma\vec{\mu}, \vec{\mu})} = \Lambda_1^{*q_1} \dots \Lambda_m^{*q_m} \mathcal{U}_0.$$

can be considered as a multidimensional generalization of Hermite functions. Indeed, in deriving the orthogonality relations and proving that the eigenfunctions of the operator (5.27) coincide with $\mathcal{U}_{q_1 \cdots q_m}$, we used only the commutation relations (5.4) and (5.22), which in turn follow from formulas (4.27) and (4.28). It is easily shown that any symmetric matrix Γ with a positive-definite imaginary part can be represented in the form $\Gamma = Z' Z^{-1}$, where Z' and Z are matrices satisfying relations (4.27) and (4.28). (It is not necessary to assume here that Z' is the derivative of Z with respect to some parameter.)

Many workers in the mathematical theory of quantum mechanics often encounter functions analogous to $\mathcal{U}_{q_1 \cdots q_m}$ (cf., for example, [14]).

LITERATURE CITED

1. Vainshtein, L. A., "Radiation flux in the triaxial ellipsoid," in High-Power Electronics [in Russian], Vol. 4, Nauka, Moscow (1965), pp. 93-106.
2. Bykov, V. P., "The geometrical optics of three-dimensional vibrations in open resonators," in: High-Power Electronics [in Russian], Vol. 4, Nauka, Moscow (1965), pp. 66-93.
3. Leontovich, M. A., "On the method of solving problems concerning the propagation of waves along the surface of the earth," Izv. Akad. Nauk SSSR, ser. fiz., Vol. B, No. 16 (1944).
4. Fok, V. A., "The field of a plane wave near the surface of a conducting body," Izv. Akad. Nauk SSSR, ser. fiz., Vol. 10, No. 2, pp, 171-186 (1946).

5. Buldyrev, V. S., "The short wave asymptotic behavior of the eigenfunctions of the Helmholtz operator," Dokl. Akad. Nauk SSSR, 163, No. 4, pp. 853-856 (1965).
6. Babich, V. M., and Lazutkin, V. F., "Eigenfunctions concentrated near a closed geodesic," in: Topics in Mathematical Physics, Vol. 2, M. Sh. Birman, ed.; Consultants Bureau, New York (1968).
7. Alekseev, A. S., Babich, V. M., and Fel'chinskii, V. Ya., "The ray method for computing the intensity of wave fronts," in: Problems in the Dynamical Theory of Propagation of Seismic Waves [in Russian], Len. Gos. Univ., No. 5, pp. 3-35 (1961).
8. Rashevskii, P. K., Riemannian Geometry and Tensor Analysis [in Russian], Mir, Moscow (1967).
9. Bishop, R. L., and Crittenden, R. J., Geometry of Manifolds, Academic Press, New York (1964).
10. Milnor, J., Morse Theory, Princeton Univ. Press, Princeton (1963).
11. Smirnov, V. I., A Course in Higher Mathematics [in Russian], Moscow (1949).
12. Bourbaki, N., Functions of a Real Variable, Elementary Theory [Russian translation], Nauka, Moscow (1965).
13. Segal, I. E., "Foundations of the theory of dynamical systems of infinitely many degrees of freedom, II," Canadian J. Math., Vol. 13, No. 1, (1961).
14. Chernikov, P. A., "A system with Hamiltonian in the form of a time-dependent quadratic form of p and q," Zh. Éksp. Teor. Fiz., Vol. 53, No. 3, pp. 1006-1017 (1967).

THE CALCULATION OF INTERFERENCE WAVES FOR DIFFRACTION BY A CYLINDER AND A SPHERE

A. I. Lanin

The wave fields for diffraction by a transparent cylinder and an elastic sphere have been studied in [1]-[4]. The principal concern of these papers was the study of waves of interference type (head waves and interference-surface waves). To describe the interference waves formulas were obtained which contain the special functions $G_M(\gamma)$ and $\Gamma_M(\gamma)$ having the form

$$G_M(\gamma)=\int_{2T_- e^{-\frac{2}{3}\pi i}}^{2T_+ e^{i\frac{\pi}{3}}} \frac{e^{i\gamma T}}{W_2(T)W_3(T)}\cdot\left[\frac{W_1(T)}{W_2(T)}\right]^M dT, \tag{1a}$$

$$\Gamma_M(\gamma)=\int_{2T_- e^{-\frac{2}{3}\pi i}}^{2T_+ e^{i\frac{\pi}{3}}} \frac{e^{i\gamma T}}{W_2'(T)W_3'(T)}\cdot\left[\frac{W_1'(T)}{W_2'(T)}\right]^M dT, \tag{1b}$$

$$T_+\sim 2;\quad T_-\sim 4,$$

where $W_1(T)$ and $W_2(T)$ are the well-known Airy functions and γ is the reduced angular distance.

In the problem of diffraction by a transparent cylinder (sphere) of radius $r=a$ the quantity $\gamma=(\vartheta-\vartheta_0)\left(\frac{k_2 a}{2}\right)^{\frac{1}{3}}$ characterizes the distance between the ray of total reflection

$$\vartheta=\vartheta_0(\rho_0,\rho,n)\equiv \arccos\frac{n}{\rho_0}+\arccos\frac{n}{\rho}-2\arccos n$$

and the point of observation ϑ; here $\rho_0=\frac{r_0}{a}$ and $\rho=\frac{r}{a}$ are dimensionless coordinates of the source and and point of observation, $n<1$ is the relative index of refraction, and k_2 is the wave number in the inner medium. In the problem of vibrations of the elastic sphere the quantity $\gamma=\left(\frac{ka}{2}\right)^{\frac{1}{3}}$ determines the angular separation of the source of vibrations on the surface of the sphere for $\vartheta=0$ to the point of observation with angular coordinate ϑ, also on the surface of the sphere (k is the wave number). The quantity M in (1), which determines the number of waves issuing from the interference process, is given by

$$M=\left|\frac{\gamma}{2\sqrt{2T_0}}\right|,\qquad 2T_0\sim 2.5-4. \tag{2}$$

Using the functions (1), it is also possible to describe the interference wave fields in more complicated diffraction problems (in problems of wave diffraction by smooth convex bodies with arbitrary radius of curvature [5]).

In the present paper formulas for the real and imaginary parts of the functions $G_M(\gamma)$ and $\Gamma_M(\gamma)$ are given, and tables for these functions have been made with a computer. The intensity of waves of interference type and of noninterfering waves of geometrical optics is calculated. Calculations are also made which characterize the behavior of the field in a neighborhood of a limiting ray for the case of diffraction of a spherical wave by a transparent sphere. The results of these computations are compared with the value of the field for the planar case.

In the problem of diffraction by a cylinder and sphere the part of the wave field corresponding to noninterfering refracted waves of geometrical optics and to waves of interference type passing through an angular distance less than π, is represented in the form

$$\sum_{m=0}^{M-1} S^{+}_{om} + S^{+}_{OM}. \tag{3}$$

The terms in this expression are contour integrals in the complex plane of the index of cylindrical functions. The following formulas have been obtained for describing noninterfering waves of geometrical optics S^{+}_{om}, $m=0,1,\ldots M-1$, and interference waves S^{+}_{OM} as a result of studying these integrals.

a) For the problem of diffraction by a transparent cylinder [2] we find for refracted waves:

$$S^{+}_{om} = \frac{e^{i\frac{\pi}{4}}}{2\sqrt{2\pi}} \frac{K_1}{\sqrt{\kappa_1 a I_{m+1}}} (iK)^m e^{i[\kappa_1(\ell_0+\ell_1)+\kappa_2(m+1)\ell_x]} \cdot \{1+\Omega\}, \tag{4}$$

where $\Omega \sim \frac{m+1}{(\kappa_2 a)^{\frac{2}{3}}} + \frac{1}{24(m+1)T_0^{\frac{3}{2}}}$, ℓ_0 is the length of the incident ray, ℓ_x is the length of one of the chords along which the refracted wave propagates inside the cylinder, ℓ_1 is the length of the ray arriving at the point of observation, κ_1 is the wave number of the exterior medium, I_{m+1} is the geometrical divergence of the ray packet for a wave passing inside the cylinder along $m+1$ chords, K_1 is the product of the coefficient of refraction for a wave propagating from the exterior medium into the interior medium and the coefficient of refraction for a wave propagating in the reverse direction, and K is the coefficient of reflection for a wave incident on the boundary from the interior region.

For an interference head wave we find:

$$S^{+}_{OM} = \frac{1}{\pi} e^{i(M+\frac{1}{2})\pi} \frac{n}{\cos\Psi_0} \frac{e^{i[\kappa_1(\ell_0+\ell_1)+\kappa_2 a(\vartheta-\vartheta_0)]}}{[\kappa_1^2(\ell_0'+a\cos\Psi_0)(\ell_1'+a\cos\Psi_0)]^{\frac{1}{2}}} \cdot \left\{G_M(\gamma)+O\left[\left(\frac{2}{\kappa_2 a}\right)^{\frac{1}{3}}+\frac{const}{\gamma}e^{-3.4\gamma}\right]\right\}, \tag{5}$$

where ℓ_0' and ℓ_1' respectively denote the length of the ray incident on the surface of the cylinder and reflected from it at an angle of total reflection Ψ_0; and the function $G_M(\gamma)$ is defined by formula (1a).

b) For the problem of excitation of an elastic homogeneous sphere [4] we find for waves of geometrical optics:

$$S^{+}_{om} = i\frac{\kappa}{2\pi\mu a^{\frac{1}{2}}} \frac{\sin^{\frac{3}{2}}\chi_m}{\sqrt{\ell_m \sin\vartheta}} \cdot R_m(\chi_m) e^{i[\kappa\ell_m - m\frac{\pi}{2}]} \cdot \{1+\Omega\}, \tag{6}$$

where χ_m is the angle at the source between the radius of the sphere and the ray arriving at the point of observation after m reflections, ℓ_m is the length of the wave path, $\sin^{\frac{3}{2}}\chi_m$ is the direction coefficient of the source, $m\frac{\pi}{2}$ is the change in phase of the wave after m-fold passage through the caustic, $R_m(\chi_m)$ is the coefficient of reflection of the wave passing along $m+1$ chords, and μ is the elastic constant.

For an interference surface wave we have:

$$S^{+}_{OM} = \frac{e^{i\pi(M-\frac{1}{4})}}{2\pi^{\frac{3}{2}}a^{2}\mu\sqrt{\sin\vartheta}} \left(\frac{\kappa a}{2}\right)^{\frac{7}{6}} e^{i\kappa a\vartheta} \cdot \left\{\Gamma_{M}(\gamma) + O\left[\left(\frac{2}{\kappa a}\right)^{\frac{2}{3}} + \frac{const}{\gamma} e^{-3.4\gamma}\right]\right\}, \tag{7}$$

where the function $\Gamma_M(\gamma)$ is given in (1b).

The intensity of the waves S^{+}_{om} and S^{+}_{OM} as a function of the reduced angular distance has been computed on the basis of formulas (4)-(7). In the computations the quantity $M \geqslant 0$ was taken to be $M = [\frac{\gamma}{3.4}]$.

1. In problem a) the computation of the intensity of the refracted waves of geometrical optics $|S^{+}_{om}|$ was made with the following values of the parameters: $\rho_0 = 2$; $\rho = 1.5$; $n = 0.5$; $(\frac{\kappa_2 a}{2})^{\frac{1}{3}} = 10$ in the interval $3.4 \leqslant \gamma \leqslant 14$. Figure 1 shows intensity curves for separated waves. For $\gamma = 3.4$ the interference process in a neighborhood of a head wave gives rise to a refracted wave of geometrical optics which passes along a single chord inside the cylinder. For $\gamma = 6.8$ the wave passing inside the cylinder along two chords takes on a distinct character, etc. Figure 1 implies that the intensity of waves passing inside the cylinder along $m+1$ chords, $m = 0, 1, \ldots M-1$, increases monotonically with γ and reaches a certain magnitude. Moreover, the intensities of waves passing over two and more chords for large γ have an effect on the order of smaller intensity of a wave passing along a single chord inside the cylinder. The decrease in intensity of multiple waves as compared with the intensity of a wave undergoing $m = 0$ reflections is explained by the fact that with an increase in the number m of reflections the geometrical divergence $I_{m+1} \sim \frac{m+1}{\cos\alpha_{om}}$ becomes larger (with increases in m the angle of refraction of the ray $\alpha_{om} \to \frac{\pi}{2}$ and $I_{m+1} \to \infty$, while $|S^{+}_{om}| \sim \frac{1}{\sqrt{I_{m+1}}}$). We note that as the point of observation moves away from the surface of the cylinder (for fixed γ) the intensity of the "split-off" waves $|S^{+}_{om}|$ decreases as $\rho^{-\frac{1}{2}}$.

The intensity of a head wave $|S^{+}_{OM}|$ is described by means of the special function $G_M(\gamma)$.

The function $G_M(\gamma)$ was tabulated on a computed (see the Appendix). Table 1 at the end of the paper gives the values of $Re\, G_M(\gamma)$, $Im\, G_M(\gamma)$, $|G_M(\gamma)|$, and $arg\, G_M(\gamma)$ for γ lying in the interval $1 \leqslant \gamma \leqslant 12$ with an increment $\Delta\gamma = 0.1$ $(M = 0, 1, 2.)$. Using this table, it is possible to demonstrate the dependence of the modulus of the function $|G_M(\gamma)|$ which characterizes the intensity of the head wave $|S^{+}_{OM}|$ for $(\frac{2}{\kappa_2 a})^{\frac{1}{3}} \ll 1$, on the reduced angular distance γ (Fig. 2).

A head wave is computed for $\gamma \geqslant 1$. With increasing γ the intensity of the head wave rises and reaches a maximum for $\gamma \approx 3.4$. In the interval $1 \leqslant \gamma \leqslant 3.4$ there are no separate refracted waves ($M = 0$).

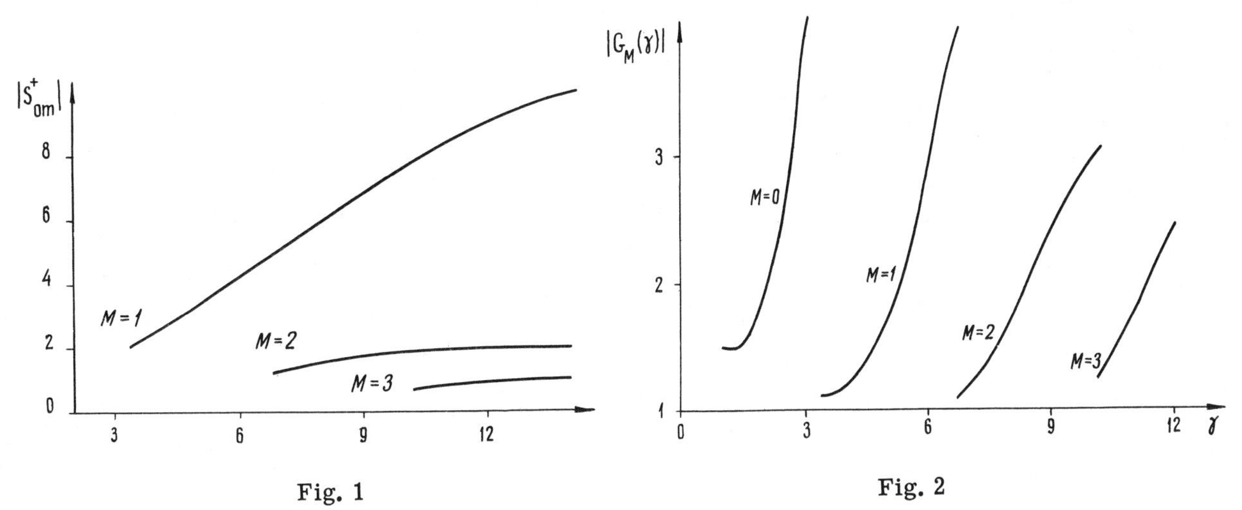

Fig. 1

Fig. 2

For $\gamma \approx 3.4$ the first separate wave emerges from the process of interference in a neighborhood of the head wave. (The number M in the formulas for calculating the noninterfering waves and the head wave is now assumed equal to one.) The intensity of the head wave computed for $\gamma = 3.4$ $(M=1)$ is found to be small, since a wave of geometrical optics has been formed from the process of interference. With increasing γ the intensity of the head wave again increases and reaches a maximum value (which is less, however, then in the case $M=0$) for $\gamma \approx 6.8$. At this value of γ a second refracted wave is separated from the interference of the head wave ($M=2$); as a result, the intensity of the head wave decreases. The process of "splitting off" waves just described is repeated as the reduced angular distance γ is further increased.

2. For problem b) the computation of the intensity of the waves of geometrical optics $|\overset{+}{\mathcal{S}}_{om}|$ and of the interference-surface wave $|\overset{+}{S}_{OM}|$ was carried out for $f=1$, H_z, $a=64 \cdot 10^7$ cm, and a transverse wave speed of $b=8 \cdot 10^5$ cm/sec.

Figure 3 shows the graphs of the intensities $|\overset{+}{\mathcal{S}}_{om}|$ of three waves depending on the reduced angular distance γ. The first wave of geometrical optics is formed for $\gamma \approx 3.4$, the second for $\gamma \approx 6.8$, etc. As γ increases, the intensities of the waves decrease monotonically, approaching some constant magnitude. The intensities of different waves are of the same order. We note that for fixed γ the intensities of the multiple waves exceed the intensity of the wave passing inside the sphere along a single chord.

The intensity of the interference surface wave is described by means of the special function $\Gamma_M(\gamma)$.

The program for computing the function $\Gamma_m(\gamma)$ is similar to that used for computing the function $G_M(\gamma)$. Table 2 for the values of the function $\Gamma_M(\gamma)$ is given for $1 \leqslant \gamma \leqslant 12.5$ with an increment of $\Delta\gamma = 0.1$.

Figure 4 shows the intensity curves $|\overset{+}{S}_{OM}|$. For changing γ in the interval $1 \leqslant \gamma \leqslant 3.4$ the intensity of the interference wave increases and reaches a maximum value at $\gamma \approx 3.4$. At $\gamma \approx 3.4$ the first wave of geometrical optics separates from the interference surface wave. As a result, the intensity of the interference wave decreases with a jump discontinuity. With further increase in γ the process of "splitting off" waves is repeated, i.e., it is the same sort of process as in the problem of diffraction by the cylinder.

As a result of splitting off waves of geometrical optics, the intensity of the interference wave experiences a jump (Figs. 2 and 4). The magnitude of this jump must be equal to the intensity of the "split-off" wave, i.e., the following relations must be satisfied:

$$|\overset{+}{S}_{00} - \overset{+}{S}_{01}| = |\overset{+}{\mathcal{S}}_{00}| \quad \text{for} \quad \gamma = 3.4, \tag{8a}$$

$$|\overset{+}{S}_{01} - \overset{+}{S}_{02}| = |\overset{+}{\mathcal{S}}_{01}| \quad \text{for} \quad \gamma = 6.8, \tag{8b}$$

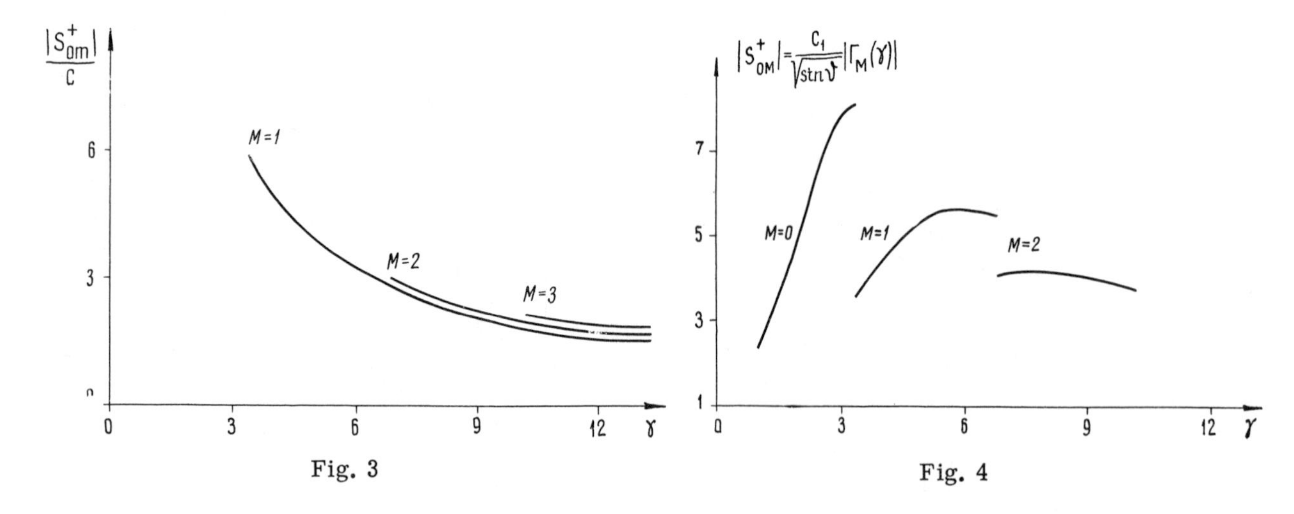

Fig. 3 Fig. 4

etc. Equations (8a) and (8b) can be used as an indirect check of the accuracy of the computations of interference waves and the waves of geometrical optics. In checking relations (8a) and (8b), the values of the functions $G_M(\gamma)$ and $\Gamma_M(\gamma)$ were used.

In the problem of diffraction by a cylinder the function $G_M(\gamma)$ describes the field of a head wave with a relative error of $\varepsilon \sim (M+1)\left(\frac{2}{k_1 a}\right)^{\frac{1}{3}}$. The relations (8) for this problem were checked with $\left(\frac{2}{k_2 a}\right)^{\frac{1}{3}} = 10^{-1}$ and $\left(\frac{2}{k_2 a}\right)^{\frac{1}{3}} = 10^{-2}$ $(n = 0.5;\ \rho = 1.5;\ \rho_0 = 2)$. The results of the computations are given below:

$\gamma = 3.4$

$\left(\frac{2}{k_2 a}\right)^{\frac{1}{3}}$	$\lvert S^+_{00} - S^+_{01} \rvert$	$\lvert S^+_{00} \rvert$	$\frac{\lvert S^+_{00} - S^+_{01}\rvert - \lvert S^+_{00}\rvert}{\lvert S^+_{00} - S^+_{01}\rvert} \cdot 100\%$
10^{-1}	0.05	0.044	11.8%
10^{-2}	$1.6327 \cdot 10^{-3}$	$1.6 \cdot 10^{-3}$	1.8%

$\gamma = 6.8$

$\left(\frac{2}{k_2 a}\right)^{\frac{1}{3}}$	$\lvert S^+_{02} - S^+_{01} \rvert$	$\lvert S^+_{01} \rvert$	$\frac{\lvert S^+_{02} - S^+_{01}\rvert - \lvert S^+_{01}\rvert}{\lvert S^+_{02} - S^+_{01}\rvert} \cdot 100\%$
10^{-1}	0.03436	0.026	24%
10^{-2}	$1.148 \cdot 10^{-3}$	$1.114 \cdot 10^{3}$	3%

These values of relative error agree for $M = 0.1$ with an error of order $\varepsilon \sim (M+1)\left(\frac{2}{k_2 a}\right)^{\frac{1}{3}}$. It can be shown that with a guaranteed degree of accuracy the formulas for describing the head wave and noninterfering waves are continuous for $\gamma = 10.2$, etc.

A similar check of the accuracy of computing the waves was made for the problem of diffraction by the elastic sphere.

3. A comparison of the wave fields of interference type in the problems of wave diffraction on a curvilinear and planar boundary is of special interest. In particular, it is of interest to compare the behavior of the interference field in a neighborhood of a ray reflected at an angle of total reflection ψ_0 in the case of a planar [6] and a spherical boundary separating two media. With this in mind, the interference field was computed in a neighborhood of a ray of total reflection for the problem of diffraction of a spherical wave by a transparent sphere with a relative index of refraction of $n = \frac{k_2}{k_1} < 1$ (k_1 and k_2 are the wave numbers for the exterior and interior of the sphere*).

The process of formation of waves of geometrical optics and of the head wave of interference type, and also the methods of computing these waves in the case of diffraction by the sphere, are the same as for the case of diffraction by the transparent cylinder.

On the basis of the results of [3], we shall write down the expression for the wave field $u(r, \vartheta)$ in the case of a sphere only in a neighborhood of the ray of total reflection ($|\gamma| = |(\vartheta - \vartheta_0)(\frac{k_2 a}{2})^{\frac{1}{3}}| \leq 1$)†

*The equation of the surface of the sphere in spherical coordinates is $(r, \vartheta, \varphi) - r = a$; $(r = r_0 > a,\ \vartheta = 0)$ are the coordinates of the source point which emits a spherical wave $\frac{e^{ik_1 R}}{R}$ (R is the distance between the source and the point of observation).

†Waves passing through an angular distance greater than π, are not considered.

$$u(r,\vartheta)\Big|_{|\gamma|<1}=\frac{e^{ik_1R}}{R}\left[1+O\left(\frac{1}{k_1R}\right)\right]+u_\Sigma(r,\vartheta)\cdot\left\{1+O\left[\left(\frac{2}{p}\right)^{\frac{2}{3}}+\frac{\text{const}}{(3,4+|\gamma|)}\cdot e^{-\frac{\sqrt{3}}{2}(1+|\gamma|)T_j}\right]\right\}, \tag{9}$$

$$p\equiv k_2a,\qquad j=1,2,$$

where the function $u_\Sigma(r,\vartheta)$, given by

$$u_\Sigma(r,\vartheta)=e^{i\frac{\pi}{4}}\frac{n}{2\sqrt{\pi}}\left(\frac{2}{p}\right)^{\frac{1}{6}}\frac{a}{\sqrt{rr_0\sin\vartheta}}\frac{e^{ik_2\left[\frac{l_0'+l_1'}{n}+a(\vartheta_0-\vartheta_1)\right]}}{[(l_0'+a\cos\psi_0)(l_1'+a\cos\psi_0)]^{\frac{1}{2}}}\cdot\left[G_1^0(\gamma,p)+G_2^0(\gamma,p)+\alpha_0G_3^0(\gamma,p)\right],$$

$$\alpha_0=4\frac{n}{\cos\psi_0}\cdot\left(\frac{2}{p}\right)^{\frac{1}{3}},$$

describes the reflected wave for $-1<\gamma<0$ and the total disturbance of the reflected and head wave for $0\leqslant\gamma\leqslant1$. The symbols l_0' and l_1' carry the meaning of the length of the ray incident on the surface of the sphere (l_0') and reflected from it (l_1') at an angle of total reflection ψ_0. The special functions $G_j^0(\gamma,p)$, $j=1,2,3$, are given by

$$G_1^0=-\int_0^{T_1e^{\frac{2}{3}\pi i}}\left[1+2i\rho\frac{w_2'(T)}{w_2(T)}\right]e^{i\gamma T-i\alpha_1T^2}dT, \tag{10a}$$

$$G_2^0=\int_0^{T_2e^{-\frac{\pi}{3}i}}\left[1+2i\rho\frac{w_3'(T)}{w_3(T)}\right]\cdot e^{i\gamma T-i\alpha_1T^2}dT, \tag{10b}$$

$$G_3^0=-\int_0^{T_3e^{-\frac{2}{3}\pi i}}\frac{e^{i\gamma T-i\alpha_1T^2}}{[w_2(T)-i\rho w_2'(T)][w_3(T)-i\rho w_3'(T)]}dT, \tag{10c}$$

$$T_3\sim5\text{-}6,$$

where $w_1(T)$ and $w_2(T)$ are Airy functions,

$$\rho=\frac{n}{\sqrt{1-n^2}}\left(\frac{2}{p}\right)^{\frac{1}{3}},$$

$$\alpha_1=\frac{n\Lambda}{4}\left(\frac{2}{p}\right)^{\frac{1}{3}},$$

$$\Lambda=-\left[\frac{a}{\sqrt{r_0^2-(na)^2}}+\frac{a}{\sqrt{r^2-(na)^2}}-\frac{2}{\sqrt{1-n^2}}\right].$$

The limits of integration T_j, $j=1,2$, in the integrals (10a, b) are given by

$$T_j=\frac{2}{n\Lambda}\left(\frac{p}{2}\right)^{\frac{1}{3}}\left[(-1)^{j+2}\gamma+\sqrt{\gamma^2+4n\Lambda\left(\frac{2}{p}\right)^{\frac{1}{3}}}\right].$$

We note that for $\left(\frac{p}{2}\right)^{\frac{1}{3}}\gg1$ in the case of a sphere the following approximation of geometrical optics can be obtained for the field in a neighborhood of the ray of total reflection:

$$u(r,\vartheta)\Big|_{|\gamma|<1} = \frac{e^{ik_1R}}{R}\left\{1+O\left(\frac{1}{k_1R}\right)\right\}+\frac{\sqrt{\sin\psi}}{\sqrt{rr_0\sin\vartheta}}\frac{e^{ik_1\ell_\Sigma}}{\sqrt{I_0}}\left\{1+O\left[\left(\frac{2}{\rho}\right)^{\frac{1}{3}}\right]\right\},$$

where ℓ_Σ is the total length of the incident and reflected ray, I_0 is the geometrical divergence of the ray packet corresponding to the reflected ray,

$$I_0=\frac{2\sqrt{r_0^2-a^2\sin^2\psi}\sqrt{r^2-a^2\sin^2\psi}}{a\cos\psi}-\frac{1}{a}\left(\sqrt{r_0^2-a^2\sin^2\psi}-\sqrt{r^2-a^2\sin^2\psi}\right)$$

and ψ is the angle of incidence of the wave on the surface of the sphere.

Suppose that at the point $Q(r_0,\vartheta=0)$ (Fig. 5) there is a source which emits a spherical wave, and suppose that the point of observation $S(r,\vartheta=\vartheta_0)$ is located on the ray reflected from the surface of the sphere at an angle of total reflection $\psi_0(\sin\psi_0=n)$. The intensity of the wave field $|u_\Sigma(S)|$ was computed on the limiting ray as a function of the radius a of the sphere. To calculate the function $|u_\Sigma(S,a)|$ a program was written for the EVM-BESM-2 computer (Appendix). The results of the computations are shown in Fig. 6 for $n=0.6$ and in Fig. 7 for $n=0.8$. The heavy curves denote the value of $|u_\Sigma(S,a)|$ computed by formula (9).*

The dashed lines denote the value of the field $|u_{pl}(S)|$ at the point S for the case of diffraction of a spherical wave by a planar boundary TT (Fig. 5) between two media with the same relative index of refraction.

The function $|u_{pl}(S)|$ can be computed from the formula

$$|u_{pl}(S)|=\frac{1}{\ell_0'+\ell_1'}\left|1-\frac{1.644\cdot tg^{\frac{1}{4}}\psi_0}{[k_1(\ell_0'+\ell_1')]^{\frac{1}{4}}}\cdot e^{i\frac{\pi}{8}}\right|,$$

in which the first term gives the approximation of geometrical optics for the reflected wave and the second term, which represents a correction to geometrical optics, takes account of the interference of the reflected wave and the ordinary head wave of noninterference type [6].

It is evident from Figs. 6 and 7 that for a fixed length of optical path $\ell_0'+\ell_1'$ the value of the field $|u_\Sigma(S,a)|$ at the point S for the sphere approaches the value of the field $|u_{pl}(S)|$ for increasing a.

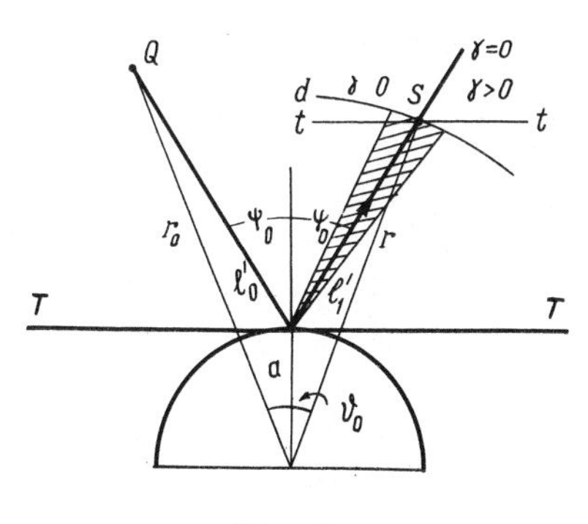

Fig. 5

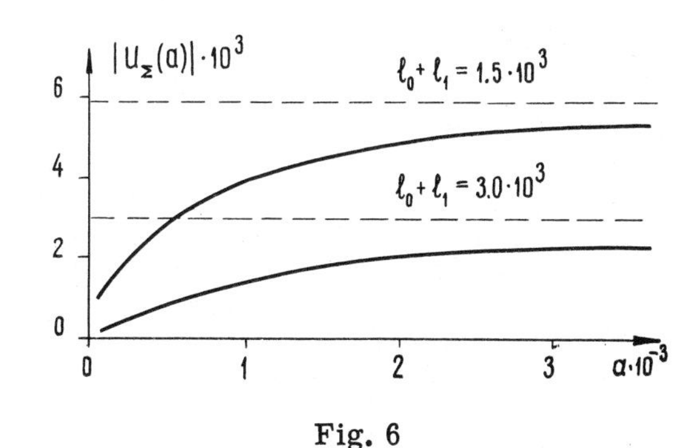

Fig. 6

*In deriving the formula describing the wave field in a neighborhood of a limiting ray for the sphere it was assumed that the point of observation and the location of the source were situated outside the sphere, i.e., that $\rho_0=\frac{r_0}{a}>\rho=\frac{r}{a}>1$. Therefore, numerical computations for a fixed length of optical path $\ell_0'+\ell_1'$ (Figure 5) were carried out for $\rho_0=1+\Delta_0$ and $\rho=1+\Delta$ where $\Delta_0\sim\Delta\gg\left(\frac{2}{\rho}\right)^{\frac{1}{3}}$. The case in which for $a\to\infty$ the values $\rho_0,\rho\to1$ (the source point and point of observation approach the surface of the sphere) must be considered separately.

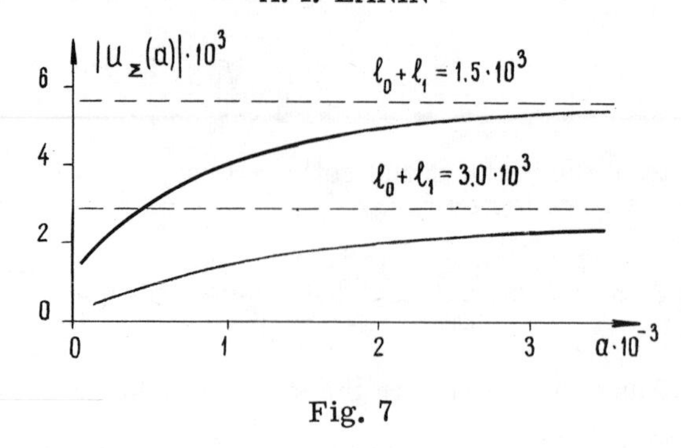

Fig. 7

We use graphs to show the results of computing the intensity of the field $|u_z(\gamma)|$ in a neighborhood of the ray of total reflection $\vartheta=\vartheta_0(r,r_0,n)$ $(|\gamma|\leq 0{,}7\text{-}1)$ for a fixed value of a. The program written for the function $|u_z(\gamma)|$ can be used to compute the function $|u_z(S,a)|$. Figures 8a and 8b show the values $|u_z(\gamma)|$ for $|\gamma|\leq 0{,}7$ $(r=\text{const})$; the parameters of the problem are shown on the figures.

The qualitative picture of the variation of the field in a neighborhood of a limiting ray for the sphere is in agreement with the results obtained for the planar case in [8]. In that paper the diffraction of a spherical wave by a planar boundary between two media with $n<1$ is studied, and the coefficient of reflection of the spherical wave is computed for various values of the parameters of the problem. For diffraction by a sphere, just as in the planar case, the field intensity reaches a maximum not on the limiting ray (for $\gamma=0$) but at a certain distance from it (beyond the critical point) $(\gamma\approx 0.7\sim 0.8)$ (Figs. 8a and 8b).

Figures 8a and b correspond to the case in which the radial coordinate of the point of observation is fixed $(r=\text{const})$, and only the angle of observation ϑ varies, i.e., a neighborhood of the limiting ray

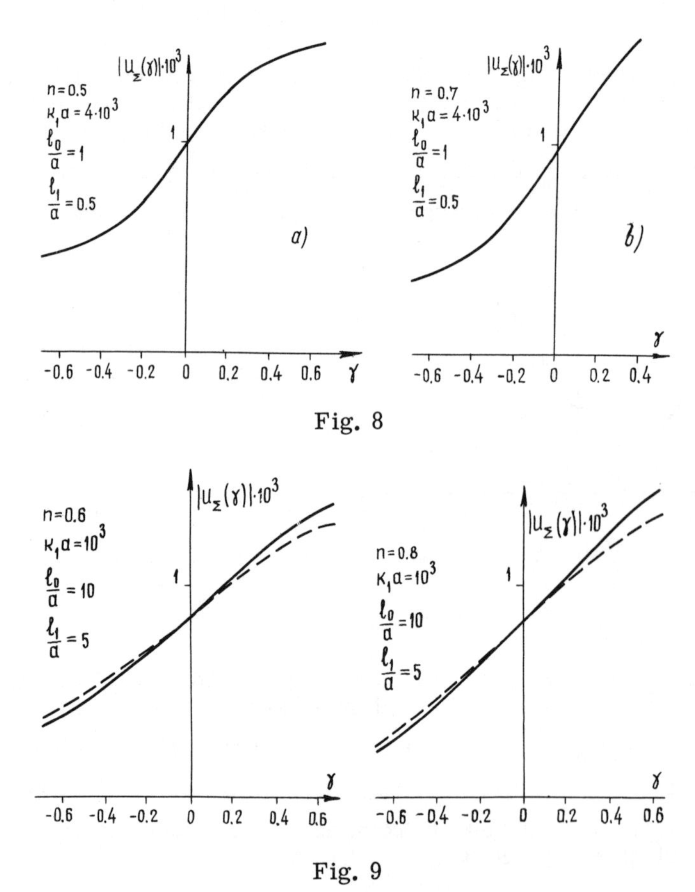

Fig. 9

is intersected along the arc dd (Fig. 5). Figure 9 shows the results of computing the field intensity $|u_{\Sigma}(\gamma)|$ (dashed lines) along the line tt (Fig. 5) parallel to the plane TT.

For comparison, Fig. 9 also shows the graphs of the field intensity $|u_{\Sigma}(\gamma)|$ (solid lines) in the case where $\tau = const$. It is evident that in the region $\gamma < 0$ the field intensity computed along the line tt exceeds that along the arc dd (in the region $\gamma > 0$).

For the case of a cylinder, the field in a neighborhood of a limiting ray can be studied in a similar manner.

The author wishes to thank V. S. Buldyrev for his constant interest in this work and for discussion of the results.

APPENDIX

Tabulation of the function $G_M(\gamma)$ in (1a) reduces to computing $\operatorname{Re} G_M(\gamma)$ and $\operatorname{Im} G_M(\gamma)$ for $\kappa_2 a \gg 1$ and values γ and M related by

$$M = \left[\frac{\gamma}{2\sqrt{2T_0}} \right], \quad 2T_0 \sim 2.5-4.$$

We replace the Airy functions $w_2(T)$ and $w_3(T)$ appearing in the integrand of the integral for $G_M(\gamma)$ by functions of $w_1(T)$ as follows:

$$\begin{aligned} w_2(T) &= e^{\frac{2}{3}\pi i} w_1(e^{\frac{2}{3}\pi i} T), \\ w_3(T) &= -\left[w_1(T) + e^{\frac{2}{3}\pi i} w_1(e^{\frac{2}{3}\pi i} T) \right], \end{aligned} \tag{I}$$

where

$$w_1(T) = \frac{1}{\sqrt{\pi}} \int_{\infty e^{-\frac{2}{3}\pi i}}^{\infty} e^{Tz - \frac{1}{3}z^3} dz.$$

We then represent $G_M(\gamma)$ as a sum of two integrals

$$G_M(\gamma) = \int_{2T_- e^{-\frac{2}{3}\pi i}}^{0} + \int_{0}^{2T_+ e^{i\frac{\pi}{3}}}.$$

The integrals over the line segments $(2T_- e^{-\frac{2}{3}\pi i}, 0)$ and $(0, 2T_+ e^{i\frac{\pi}{3}})$ can be reduced to integrals over the intervals $(2T_-, 0)$ and $(0, 2T_+)$ of the real line. If we then use the relations [9]

$$\begin{aligned} w_1(T) &= u(T) + i v(T), \\ w_1(-T) &= u(-T) + i v(-T), \\ w_1(e^{i\frac{\pi}{3}} T) &= 2 e^{i\frac{\pi}{6}} v(-T), \\ w_1(e^{-\frac{2}{3}\pi i} T) &= 2 e^{i\frac{\pi}{6}} v(T), \end{aligned} \tag{II}$$

where $u(T)$ and $v(T)$ are real Airy functions, and carry out the corresponding computations, then for $\operatorname{Re} G_M(\gamma)$ and $\operatorname{Im} G_M(\gamma)$ we obtain the following representations:

$$\mathrm{Re}\,G_M(\gamma)=\int_0^{2T}\left\{e^{X(T)}g(T)\cos\chi(T)+e^{X(-T)}g(-T)\cos\chi(-T)\right\}dT,$$
$$\mathrm{Im}\,G_M(\gamma)=-\int_0^{2T}\left\{e^{X(T)}g(T)\sin\chi(T)+e^{X(-T)}g(-T)\sin\chi(-T)\right\}dT, \tag{III}$$

where

$$X(T)=\frac{\sqrt{3}}{2}\gamma T-\ln[u^2(T)+v^2(T)]-M\ln\sqrt{u^2(T)+v^2(T)},$$

$$g(T)=[2v(T)]^M,$$

$$\chi(T)=\frac{\gamma}{2}T+M\left(\operatorname{arctg}\frac{v(T)}{u(T)}+\frac{\pi}{2}\right).$$

A program for the EVM-BESM-2 computer was written to compute the integrals (III). In addition to $\mathrm{Re}\,G_M(\gamma)$ and $\mathrm{Im}\,G_M(\gamma)$, the program included the computation of $|G_M(\gamma)|$ and $\arg G_M(\gamma)$. Expressions similar to (III) can be obtained for $\mathrm{Re}\,\Gamma_M(\gamma)$ and $\mathrm{Im}\,\Gamma_M(\gamma)$ (Ib) in which the functions $X(T)$, $g(T)$ and $\chi(T)$ have the form

$$X(T)=\frac{\sqrt{3}}{2}\gamma T-\ln[u'^2(T)+v'^2(T)]-M\ln\sqrt{u'^2(T)+v'^2(T)},$$

$$g(T)=[2v'(T)]^M,$$

$$\chi(T)=\frac{\gamma}{2}T+M\left(\operatorname{arctg}\frac{v'(T)}{u'(T)}+\frac{11}{6}\pi\right).$$

In tabulating the integrals for the real and imaginary parts of the functions $G_M(\gamma)$ and $\Gamma_M(\gamma)$, the real Airy functions $u(T)$, $v(T)$ and their derivatives $u'(T)$, $v'(T)$, which appear in the integrands, are computed by a standard program [10].

Computation of the function $|u_\Sigma(r,\vartheta)|$. For the modulus of the function $|u_\Sigma(r,\vartheta,\alpha)|$ in (9) we have

$$|u_\Sigma(r,\vartheta,\alpha)|=A_0\frac{1}{\sqrt{\sin\vartheta}}\left|\overset{\circ}{G}_1(\gamma,p)+\overset{\circ}{G}_2(\gamma,p)+\alpha_0\overset{\circ}{G}_3(\gamma,p)\right|, \tag{IV}$$

where

$$A_0=\frac{n}{2\sqrt{\pi}}\left(\frac{2}{p}\right)^{\frac{1}{6}}\frac{a}{\sqrt{rr_0}}[(l_0'+a\cos\psi_0)(l_1'+a\cos\psi_0)]^{-\frac{1}{2}}.$$

The special functions $\overset{\circ}{G}_j(\gamma,p)$ in formula (IV) are defined by equations (10).

We shall now find expressions for $\mathrm{Re}\,\overset{\circ}{G}_j(\gamma,p)$ and $\mathrm{Im}\,\overset{\circ}{G}_j(\gamma,p)$, $j=1,2,3$. We make use of formulas (I) for the Airy functions $w_{j+1}(T)$, $j=1,2$, in the integrals $\overset{\circ}{G}_j(\gamma,p)$ and reduce the integrals for $\overset{\circ}{G}_j(\gamma,p)$ extending over a line segment in the complex plane to integrals over intervals of the real line. Taking account of (II) and carrying out the computations, we obtain for $\mathrm{Re}\,\overset{\circ}{G}_j(\gamma,p)$ and $\mathrm{Im}\,\overset{\circ}{G}_j(\gamma,p)$ the expressions

$$Re\, G_1^0 = \int_0^{T_1} e^{X_1(T)} \left[\tfrac{1}{2}\cos\chi_1(T) - \left(\tfrac{\sqrt{3}}{2} + 2\beta\frac{v'(T)}{v(T)}\right)\sin\chi_1(T)\right] dT,$$

$$Im\, G_1^0 = -\int_0^{T_1} e^{X_1(T)} \left[\tfrac{1}{2}\sin\chi_1(T) + \left(\tfrac{\sqrt{3}}{2} + 2\beta\frac{v'(T)}{v(T)}\right)\cos\chi_2(T)\right] dT,$$

where

$$X_1(T) = -\frac{\sqrt{3}}{2}(\gamma + \alpha_1 T)\, T,$$

$$\chi_1(T) = (\gamma - \alpha_1 T)\,\frac{T}{2}\ ;$$

$$Re\, G_2^0 = \int_0^{T_2} e^{X_2(T)} \left[\left(\tfrac{1}{2} + 2\beta\frac{u v' - u' v}{u^2 + v^2}\right)\cos\chi_2 + \left(\tfrac{\sqrt{3}}{2} + 2\beta\frac{u \cdot u' + v \cdot v'}{u^2 + v^2}\right)\sin\chi_2\right] \cdot dT,$$

$$Im\, G_2^0 = \int_0^{T_2} e^{X_2(T)} \left[\left(\tfrac{1}{2} + 2\beta\frac{u \cdot v' - u' v}{u^2 + v^2}\right)\sin\chi_2 + \left(\tfrac{\sqrt{3}}{2} + 2\beta\frac{u \cdot u' + v \cdot v'}{u^2 + v^2}\right)\cos\chi_2\right] \cdot dT,$$

where

$$X_2(T) = \frac{\sqrt{3}}{2}(\gamma - \alpha_1 T)\, T,$$

$$\chi_2(T) = (\gamma + \alpha_1 T)\,\frac{T}{2}\ ;$$

$$Re\, G_3^0 = \int_0^{T_3} \frac{e^{X_3(T)}}{u^2(T) + v^2(T)} \cos\chi_3(T)\, dT,$$

$$Im\, G_3^0 = -\int_0^{T_3} \frac{e^{X_3(T)}}{u^2(T) + v^2(T)} \cdot \sin\chi_3(T)\, dT,$$

where

$$X_3(T) = \frac{\sqrt{3}}{2}\, T(\gamma + \alpha_1 T) - \beta\gamma,$$

$$\chi_3(T) = (\gamma - \alpha_1 T)\,\frac{T}{2}\ .$$

At the end points of the integral of integration the integrands of $Re\, G_j^0$ and $Im\, G_j^0$, $j = 1, 2, 3$, are exponentially small.

A program for computing the function $|U_z|$ was written for the BESM-2 computer. The program can be used for computing $|U_z|$ as a function of any of the parameters appearing in expression (IV) for $|U_z|$.

TABLE I

γ	$\mathrm{Re}\, G_M(\gamma)$	$\mathrm{Im}\, G_M(\gamma)$	$\lvert G_M(\gamma)\rvert$	$\arg G_M(\gamma)$
		$M = 0$		
1.0	1.641	0.222	1.656	0.134
1.1	1.579	0.046	1.580	0.029
1.2	1.527	-0.125	1.533	-0.081
1.3	1.479	-0.298	1.508	-0.1989
1.4	1.426	-0.476	1.504	-0.322
1.5	1.364	-0.662	1.516	-0.452
1.6	1.284	-0.856	1.544	-0.588
1.7	1.181	-1.059	1.586	-0.731
1.8	1.047	-1.266	1.643	-0.880
1.9	0.874	-1.475	1.715	-1.036
2.0	0.655	-1.678	1.801	-1.198
2.1	0.383	-1.865	1.904	-1.368
2.2	0.052	-2.022	2.023	-1.544
2.3	-0.340	-2.134	2.161	-1.729
2.4	-0.794	-2.177	2.317	-1.921
2.5	-1.302	-2.128	2.495	-2.120
2.6	-1.850	-1.960	2.695	-2.327
2.7	-2.411	-1.646	2.920	-2.542
2.8	-2.950	-1.161	3.170	-2.766
2.9	-3.414	-0.490	3.448	-2.999
3.0	-3.738	0.371	3.756	3.043
3.1	-3.846	1.403	4.094	2.792
3.2	-3.657	2.557	4.463	2.531
3.3	-3.095	3.749	4.861	2.261
		$M = 1$		
3.4	-1.087	-0.227	1.110	-2.935
3.5	-1.112	-0.119	1.118	-3.035
3.6	-1.129	-0.001	1.129	-3.141
3.7	-1.136	0.126	1.143	3.031
3.8	-1.130	0.262	1.160	2.914
3.9	-1.109	0.406	1.180	2.791
4.0	-1.068	0.556	1.205	2.662
4.1	-1.007	0.710	1.233	2.528
4.2	-0.922	0.865	1.265	2.388
4.3	-0.810	1.018	1.300	2.243
4.4	-0.669	1.163	1.342	2.093
4.5	-0.498	1.294	1.387	1.938
4.6	-0.296	1.407	1.437	1.778
4.7	-0.063	1.492	1.493	1.613
4.8	0.196	1.542	1.554	1.444

TABLE I (Continued)

γ	$\mathrm{Re}\,G_M(\gamma)$	$\mathrm{Im}\,G_M(\gamma)$	$\lvert G_M(\gamma)\rvert$	$\arg G_M(\gamma)$
		M=1		
4.9	0.483	I.548	I.622	I.268
5.0	0.785	I.502	I.695	I,089
5.I	I.096	I.396	I.775	0,904
5.2	I.404	I,223	I.862	0,7I6
5.3	I,695	0,977	I.956	0.523
5.4	I,950	0,657	2,058	0.325
5.5	2,I5I	0,264	2,I67	0.I22
5.6	2.276	-0,I95	2.284	-0,085
5.7	2,305	-0,707	2,4II	-0.298
5.8	2,2I7	-I,253	2.546	-0,5I5
5.9	I.990	-I,806	2,688	-0,737
6.0	I.6I9	-2.333	2.840	-0.964
6.I	I,099	-2.793	3,002	-0,I96
6.2	0,436	-3,I43	3,I72	-I.433
6,3	-0.350	-3.333	3.352	-I.675
6.4	-I.225	-3.323	3,542	-I,924
6.5	-2,I32	-3,073	3.74I	-2,I77
6.6	-3,005	-2,557	3,946	-2,436
6,7	-3.765	-I.I70	4,I6I	-2.702
		M=2		
6,8	0,759	-0,804	I,I06	-0,8I4
6,9	0,647	-0.93I	I,I34	-0,963
7,0	0,5I0	-I,048	I.I65	-I.II7
7,I	0,349	-I,I46	I,I98	-I,274
7,2	0,I66	-I,223	I,235	-I,435
7,3	-0,037	-I,273	I.273	-I.600
7.4	-0,256	-I.29I	I,3I6	-I.767
7.5	-0,489	-I,270	I.36I	-I.938
7.6	-0.726	-I.208	I.409	-2,II2
7.7	-0,962	-I,099	I.46I	-2,289
7.8	-I.I86	-0,943	I.5I6	-2.470
7.9	-I.39I	-0,739	I.575	-2,653
8.0	-I.563	-0.487	I.637	-2,839
8.I	-I.69I	-0.I9I	I.702	-3.029
8.2	-I.764	0.I40	I.770	3.062
8.3	-I.775	0.500	I.844	2.866
8.4	-I.709	0.873	I.9I9	2,669
8.5	-I.562	I.243	I.997	2.470

TABLE I (Continued)

γ	$\mathrm{Re}\, G_M(\gamma)$	$\mathrm{Im}\, G_M(\gamma)$	$\lvert G_M(\gamma)\rvert$	$\arg G_M(\gamma)$
		$M=2$		
8.6	-I.329	I.600	2.079	2.264
8.7	-I.0I2	I.9I2	2.I63	2.058
8.8	-0.6I5	2.I62	2.248	I.848
8.9	-0.I50	2.328	2.333	I.635
9.0	0.366	2.39I	2.4I9	I.4I9
9.I	0.909	2.33I	2.502	I.I99
9.2	I.448	2.I37	2.58I	0.975
9.3	I.939	I.803	2.648	0.749
9.4	2.372	I.336	2.723	0.5I3
9.5	2.678	0.750	2.78I	0.273
9.6	2.828	0.073	2.829	0.026
9.7	2.788	-0.654	2.864	-0.230
9.8	2.536	-I.379	2.886	-0.498
9.9	2.063	-2.038	2.900	-0.780
I0.0	I.376	-2.562	2.908	-I.078
I0.I	0.504	-2.878	2.922	-I.397
		$M=3$		
I0.2	0.692	I.038	I.247	0.983
I0.3	0.89I	0.932	I.289	0.808
I0.4	I.078	0.787	I.335	0.63I
I0.5	I.244	0.604	I.383	0.452
I0.6	I.383	0.385	I.436	0.272
I0.7	I.486	0.I33	I.492	0.089
I0.8	I.544	-0.I46	I.55I	-0.095
I0.9	I.552	-0.445	I.6I5	-0.279
II.0	I.506	-0.755	I.685	-0.465
II.I	I.398	-I.065	I.757	-0.65I
II.2	I.228	-I.363	I.835	-0.837
II.3	0.995	-I.636	I.9I6	-I.024
II.4	0.704	-I.872	2.000	-I.2II
II.5	0.362	-2.056	2.088	-I.396
II.6	-0.023	-2.I78	2.I78	-I.58I
II.7	-0.435	-2.226	2.268	-I.764
II.8	-0.862	-2.I95	2.359	-I.945
II.9	-I.285	-2.075	2.44I	-2.I25
I2.0	-I.700	-I.950	2.520	-2.300

TABLE II

γ	$Re\,\Gamma_M(\gamma)$	$Im\,\Gamma_M(\gamma)$	$\|\Gamma_M(\gamma)\|$	$arg\,\Gamma_M(\gamma)$
		$M=0$		
I	0.743	0.004	0.743	0.006
I.I	0.852	-0.I50	0.865	-0.I74
I.2	0.950	-0.3II	I.000	-0.3I6
I.3	I.036	-0.48I	I.I42	-0.434
I.4	I.II0	-0.662	I.29I	-0.538
I.5	I.I67	-0.856	I.448	-0.633
I.6	I.208	-I.065	I.6I0	-0.723
I.7	I.228	-I.288	I.780	-0.809
I.8	I.223	-I.526	I.956	-0.895
I.9	I.I89	-I.778	2.I38	-0.98I
2.0	I.I20	-2.04I	2.328	-I.069
2.I	I.0II	-2.3I2	2.523	-I.I58
2.2	0.857	-2.586	2.724	-I.250
2.3	0.653	-2.856	2.929	-I.346
2.4	0.393	-3.II3	3.I37	-I.445
2.5	0.075	-3.344	3.345	-I.548
2.6	-0.303	-3.538	3.55I	-I.656
2.7	-0.736	-3.677	3.750	-I.768
2.8	-I.2I9	-3.745	3.938	-I.885
2.9	-I.740	-3.724	4.II0	-2.008
3.0	-2.279	-3.599	4.260	-2.I35
3.I	-2.8I2	-3.357	4.379	-2.268
3.2	-3.306	-2.994	4.460	-2.406
3.3	-3.725	-2.5I3	4.493	-2.548
		$M=1$		
3.4	I.783	I.056	2.072	0.534
3.5	I.942	0.974	2.I73	0.465
3.6	2.I02	0.874	2.276	0.394
3.7	2.259	0.753	2.382	0.322
3.8	2.4I3	0.6II	2.489	0.248
3.9	2.559	0.445	2.598	0.I72
4.0	2.695	0.256	2.707	0.095
4.I	2.8I7	0.044	2.8I8	0.0I5
4.2	2.922	-0.I93	2.928	-0.066
4.3	3.004	-0.453	3.038	-0.I50
4.4	3.060	-0.734	3.I47	-0.235
4.5	3.085	-I.035	3.254	-0.324
4.6	3.073	-I.352	3.358	-0.4I4
4.7	3.022	-I.68I	3.458	-0.507

TABLE II (Continued)

γ	$Re\,\Gamma_M(\gamma)$	$Im\,\Gamma_M(\gamma)$	$\lvert\Gamma_M(\gamma)\rvert$	$arg\,\Gamma_M(\gamma)$
		$M=1$		
4.8	2,926	-2,0I6	3,553	-0.603
4.9	2.782	-2.35I	3.643	-0.70I
5.0	2.588	-2.680	3.726	-0.803
5.I	2.342	-2.992	3.800	-0.906
5,2	2.045	-3.280	3.865	-I.0I3
5.3	I.698	-3.53I	3.9I8	-I.I22
5.4	I.306	-3.737	3.960	-I.234
5.5	0.876	-3.890	3.986	-I.350
5.6	0.4I6	-3.976	4.000	-I.466
5.7	-0.062	-3.992	3.993	-I.586
5.8	-0.544	-3.93I	3.969	-I.708
5.9	-I.0I7	-3.79I	3.925	-I.833
6.0	-I.462	-3.574	3.86I	-I.959
6.I	-I.862	-3.284	3.775	-2.086
6.2	-2.20I	-2.93I	3.666	-2.2I5
6.3	-2.463	-2.53I	3.532	-2.342
6.4	-2.643	-2.I03	3.377	-2.470
6.5	-2.732	-I.666	3.200	-2.594
6.6	-2.733	-I.248	3.000	-2.7I3
6.7	-2.653	-0.868	2.792	-2.825
		$M=2$		
6.8	4.654	I.880	3.252	0.6I6
6.9	2.864	I.635	3.298	0.5I9
7.0	3.05I	I.358	3.339	0.4I8
7.I	3.2I0	I.052	3.378	0.3I6
7.2	3.335	0.7I9	3.4I2	0.2I2
7.3	3.423	0.362	3.442	0.I05
7.4	3.467	-0.0I2	3.467	-0.003
7.5	3.464	-0.400	3.487	-I.I48
7.6	3.4I2	-0.793	3.503	-0.228
7.7	3.306	-I.I85	3.5I3	-0.344
7.8	3.I48	-I.568	3.5I7	-0.462
7.9	2.936	-I.933	3.5I6	-0.582
8.0	2.673	-2.27I	3.508	-0.704
8.I	2.363	-2.574	3.495	-0.828
8.2	2.0II	-2.834	3.475	-0.953
8.3	I.623	-3.042	3.448	-I.080
8.4	I.209	-3.I92	3.4I3	-I.208
8.5	0.779	-3.28I	3.372	-I.337

TABLE II (Continued)

γ	$Re\,\Gamma_M(\gamma)$	$Im\,\Gamma_M(\gamma)$	$\lvert\Gamma_M(\gamma)\rvert$	$arg\,\Gamma_M(\gamma)$
		$M=2$		
8.6	0.344	-3.306	3.324	-I.467
8.7	-0.084	-3.265	3.266	-I.596
8.8	-0.493	-3.I60	3.I98	-I.725
8.9	-0.872	-3.000	3.I24	-I.853
9.0	-I.208	-2.790	3.040	-I.979
9.I	-I.498	-2.543	2.950	-2.I02
9.2	-I.725	-2.268	2.849	-2.22I
9.3	-I.899	-I.983	2.746	-2.335
9.4	-2.009	-I.700	2.632	-2.440
9.5	-2.063	-I.435	2.5I3	-2.534
9.6	-2.076	-I.202	2.399	-2.6I7
9.7	-2.068	-I.009	2.30I	-2.687
9.8	-2.035	-0.865	2.2I2	-2.739
9.9	-2.005	-0.773	2.I49	-2.773
I0.0	-I.994	-0.729	2.I23	-2.790
I0.I	-2.02I	-0.726	2.I47	-2.796
		$M=3$		
I0.2	3.224	0.567	3.273	0.I74
I0.3	3.290	0.I66	3.294	0.050
I0.4	3.305	-0.243	3.3I4	-0.073
I0.5	3.266	-0.655	3.33I	-0.I98
I0.6	3.I72	-I.06I	3.345	-0.323
I0.7	3.026	-I.452	3.357	-0.447
I0.8	2.828	-I.822	3.364	-0.572
I0.9	2.584	-2.I60	3.368	-0.696
II.0	2.297	-2.462	3.367	-0.820
II.I	I.976	-2.720	3.362	-0.942
II.2	I.627	-2.930	3.352	-I.063
II.3	I.258	-3.089	3.335	-I.I84
II.4	0.879	-3.I96	3.3I4	-I.302
II.5	0.499	-3.250	3.288	-I.4I8
II.6	0.I26	-3.253	3.255	-I.532
II.7	-0.23I	-3.2I2	3.220	-I.642
II.8	-0.567	-3.I30	3.I8I	-I.750
II.9	-0.877	-3.0I5	3.I40	-I.853
I2.0	-I.I57	-2.876	3.I00	-I.953
I2.I	-I.407	-2.72I	3.063	-2.048
I2.2	-I.632	-2.558	3.035	-2.I38
I2.3	-I.834	-2.395	3.0I7	-2.224
I2.4	-2.023	-2.236	3.0I6	-2.306
I2.5	-2.209	-2.083	3.036	-2.385

LITERATURE CITED

1. Buldyrev, V. S., "Investigation of the Green's function for the problem of diffraction by a transparent cylinder with relative index of refraction less than one. Numerical methods for solving differential and integral equations and quadrature formulas." Supplement to Zhurnal Vychislitel'noi Matematiki i Matematicheskoi Fiziki, No. 4, Nauka (1964).
2. Buldyrev, V. S., and Lanin, A. I., "Investigation of the Green's function for the problem of diffraction by a transparent circular cylinder," Supplement to Zhurnal Vychislitel'noi Matematiki i Matematicheskoi Fiziki, No. 1, Nauka (1966).
3. Lanin, A. I., "Computation of the wave field in a neighborhood of a ray of total reflection," Trud. Matem. Inst. im. V. A. Steklova Akad. Nauk SSSR, Vol. 92 (1966).
4. Buldyrev, V. S., and Lanin, A. I., "Investigation of the interference wave field on the surface of an elastic sphere," in: Numerical Methods for the Solution of Problems of Mathematical Physics [in Russian], Nauka (1966).
5. Buldyrev, V. S., "Interference of short waves in the problem of diffraction by an inhomogeneous cylinder of arbitrary cross section," Izv. Vysshikh Uchebnykh Zavedenii, Radiofizika, Vol. 10 (1967).
6. Brekhoskikh, L. N., Waves in Layered Media [in Russian], Izd. Akad. Nauk SSSR (1957).
7. Lanin, A. I., Candidate's Dissertation, Len. Gos. Univ. (1966).
8. Cerveny, V., and Hron, F., "Reflection coefficients for spherical waves," Studia geophys. et geod., Vol. 5 (1961).
9. Fok, V. A., Tables of Airy Functions [in Russian], Moscow (1946).
10. Lanin, A. I., "Computation of Airy functions and their derivatives," VTs (LOTsEMI) Akad. Nauk SSSR, Library of BESM-2 Programs.

THE EXPANSION OF AN ARBITRARY FUNCTION IN TERMS OF AN INTEGRAL OF ASSOCIATED LEGENDRE FUNCTIONS OF FIRST KIND WITH COMPLEX INDEX

B. G. Nikolaev

The solution of a number of problems in mathematical physics reduces to finding a function $\varphi_\mu(\nu)$ from the condition

$$\Psi_\mu(x) = \int_0^\infty P^{-\mu}_{i\nu-\frac{1}{2}}(x)\,\varphi_\mu(\nu)\,d\nu, \qquad (\operatorname{Re}\mu > -\tfrac{1}{2}, \quad x \geqslant 1), \tag{1}$$

where $P^q_p(x)$ is an associated Legendre function of first kind and $\Psi_\mu(x)$ is a function defined on $1 \leqslant x < \infty$. In other words, the problem reduces to inverting the integral (1) and expanding a given function $\Psi_\mu(x)$ in an integral of associated Legendre functions. The necessity of such transformations arises, for example, in problems related to the use of toroidal and ellipsoidal coordinates.

The transformation (1) is a generalization of the well-known Mehler–Fok transform [1-2] to the case $\mu \neq 0$, and its study is of independent theoretical interest.

Before proceeding to the inversion of the integral (1), we recall certain properties of the associated Legendre functions which we shall need in the sequel.

§1. Some Properties of Associated Legendre Functions with Complex Index

The integral

$$P^\mu_{i\nu-\frac{1}{2}}(\operatorname{ch}\alpha) = \frac{2}{\pi}\,\frac{(2\operatorname{sh}\alpha)^\mu\,\Gamma(\frac{1}{2})}{\Gamma(\frac{1}{2}-\mu)}\int_0^\alpha \frac{\cos\nu t\,dt}{[2(\operatorname{ch}\alpha-\operatorname{ch}t)]^{\mu+\frac{1}{2}}} \qquad (\alpha > 0, \quad \operatorname{Re}\mu < \tfrac{1}{2}) \tag{2}$$

is taken as the definition of the associated function of first kind; the function of second kind is defined by

$$Q^{\mu}_{i\nu-\frac{1}{2}}(\mathrm{ch}\,\alpha)=\frac{e^{\mu\pi i}(2\,\mathrm{sh}\,\alpha)^{\mu}\,\Gamma(\frac{1}{2})}{\Gamma(\frac{1}{2}-\mu)}\int\limits_{\alpha}^{\infty}\frac{e^{-i\nu t}\,dt}{[2(\mathrm{ch}\,t-\mathrm{ch}\,\alpha)]^{\mu+\frac{1}{2}}} \tag{3}$$

$$(\alpha>0,\quad \mathrm{Re}\,\mu<\tfrac{1}{2},\quad \mathrm{Re}(\mu+i\nu)>-\tfrac{1}{2}).$$

The associated functions of first and second kind are related by [[3], p. 197, formula (31)]

$$\mathrm{sh}\,\nu\pi\,\sin\mu\pi\,[Q^{\mu}_{-i\nu-\frac{1}{2}}(\mathrm{ch}\,\alpha)+Q^{\mu}_{i\nu-\frac{1}{2}}(\mathrm{ch}\,\alpha)]-$$

$$-i\,\mathrm{ch}\,\nu\pi\,\cos\mu\pi\,[Q^{\mu}_{-i\nu-\frac{1}{2}}(\mathrm{ch}\,\alpha)-Q^{\mu}_{i\nu-\frac{1}{2}}(\mathrm{ch}\,\alpha)]=\pi e^{\mu\pi i}\,\mathrm{sh}\,\nu\pi\,P^{\mu}_{i\nu-\frac{1}{2}}(\mathrm{ch}\,\alpha). \tag{4}$$

A comparison of the two preceding formulas gives

$$\mathrm{ch}\,\nu\pi\,\cos\mu\pi\int\limits_{\alpha}^{\infty}\frac{\sin\nu t\,dt}{(\mathrm{ch}\,t-\mathrm{ch}\,\alpha)^{\mu+\frac{1}{2}}}+\mathrm{sh}\,\nu\pi\,\sin\mu\pi\int\limits_{\alpha}^{\infty}\frac{\cos\nu t\,dt}{(\mathrm{ch}\,t-\mathrm{ch}\,\alpha)^{\mu+\frac{1}{2}}}=\sqrt{\frac{\pi}{2}}\,\mathrm{sh}\,\nu\pi\,\frac{\Gamma(\frac{1}{2}-\mu)}{\mathrm{sh}^{\mu}\alpha}\,P^{\mu}_{i\nu-\frac{1}{2}}(\mathrm{ch}\,\alpha). \tag{5}$$

Using some identities ([3], p. 262, formula (142); p. 263, Section 29, Subsection 181) and relation (5), we obtain

$$\int\limits_{\alpha}^{\infty}\frac{\sin\nu t\,dt}{(\mathrm{ch}\,t-\mathrm{ch}\,\alpha)^{\mu+\frac{1}{2}}}=\sqrt{\frac{\pi}{2}}\,\frac{\mathrm{sh}\,\nu\pi}{\cos\mu\pi}\,\frac{\Gamma(i\nu+\mu+\frac{1}{2})\,\Gamma(-i\nu+\mu+\frac{1}{2})}{\Gamma(\mu+\frac{1}{2})\,\mathrm{sh}^{\mu}\alpha}\,P^{-\mu}_{i\nu-\frac{1}{2}}(\mathrm{ch}\,\alpha) \tag{6}$$

$$(\mathrm{Re}\,\mu<\tfrac{1}{2},\quad \mathrm{Re}(\mu\pm i\nu)>-\tfrac{1}{2}).$$

Substituting the expression

$$\left[\frac{\mathrm{sh}\,\alpha}{2(\mathrm{ch}\,\alpha-\mathrm{ch}\,t)}\right]^{\mu+\frac{1}{2}}=\left[\frac{\alpha}{\alpha^{2}-t^{2}}\right]^{\mu+\frac{1}{2}}\sum_{n=0}^{\infty}b_{n\mu}(\alpha)(\alpha^{2}-t^{2})^{n}, \tag{7}$$

$$b_{0\mu}(\alpha)=1,\quad b_{1\mu}(\alpha)=(\tfrac{1}{2}+\mu)\,\frac{\mathrm{cth}\,\alpha-\frac{1}{\alpha}}{4\alpha},\ \ldots$$

in powers (α^2-t^2) into (2) and integrating termwise using the identity

$$\int\limits_{0}^{\alpha}(\alpha^{2}-t^{2})^{\nu-\frac{1}{2}}\cos\tau t\,dt=\sqrt{\pi}\,2^{\nu-1}\,\Gamma(\nu+\tfrac{1}{2})\left(\frac{\alpha}{\tau}\right)^{\nu}J_{\nu}(\alpha\tau) \tag{8}$$

$$(\mathrm{Re}\,\nu>-\tfrac{1}{2})$$

we obtain

$$P^{-\mu}_{i\nu-\frac{1}{2}}(\mathrm{ch}\,\alpha)=\sqrt{\frac{\alpha}{\mathrm{sh}\,\alpha}}\sum_{n=0}^{\infty}c_{n\mu}(\alpha)\,\frac{J_{n+\mu}(\nu\alpha)}{\nu^{n+\mu}}, \tag{9}$$

where $c_{0\mu}(\alpha)=1,\quad c_{1\mu}(\alpha)=(\frac{1}{4}-\mu^{2})\dfrac{\mathrm{cth}\,\alpha-\frac{1}{\alpha}}{2},\ \ldots$

For $\nu\gg1$ equation (9) may be considered an asymptotic expansion which is valid for all $\alpha\geqslant0$. The corresponding series for associated Legendre functions of second kind have the form

$$Q^{-\mu}_{i\nu-\frac{1}{2}}(\operatorname{ch}\alpha)=-\frac{\pi i}{2}\sqrt{\frac{\alpha}{\operatorname{sh}\alpha}}\sum_{n=0}^{\infty}c_{n\mu}(\alpha)\frac{H^{(2)}_{n+\mu}(\nu\alpha)}{\nu^{n+\mu}},$$

$$Q^{-\mu}_{-i\nu-\frac{1}{2}}(\operatorname{ch}\alpha)=\frac{\pi i}{2}\sqrt{\frac{\alpha}{\operatorname{sh}\alpha}}\sum_{n=0}^{\infty}c_{n\mu}(\alpha)\frac{H^{(1)}_{n+\mu}(\nu\alpha)}{\nu^{n+\mu}}, \tag{10}$$

where the $H^{(\nu)}_{p}(z)$ are Hankel functions and the $c_{n\mu}(\alpha)$ take the same values as in formula (9). Formulas (10) are obtained from (3) by a similar technique, using the identity

$$\int_{\alpha}^{\infty}\frac{e^{i\tau t}}{(t^2-\alpha^2)^{\nu+\frac{1}{2}}}\,dt=i\sqrt{\pi}\,2^{-\nu-1}\left(\frac{\tau}{\alpha}\right)^{\nu}\Gamma\left(\tfrac{1}{2}-\nu\right)H^{(1)}_{\nu}(\alpha\tau),$$

$$\int_{\alpha}^{\infty}\frac{e^{-i\tau t}}{(t^2-\alpha^2)^{\nu+\frac{1}{2}}}\,dt=-i\sqrt{\pi}\,2^{-\nu-1}\left(\frac{\tau}{\alpha}\right)^{\nu}\Gamma\left(\tfrac{1}{2}-\nu\right)H^{(2)}_{\nu}(\alpha\tau), \tag{11}$$

$$(|\operatorname{Re}\nu|<\tfrac{1}{2}),$$

and also the relations

$$H^{(1)}_{-\nu}(z)=e^{i\pi\nu}H^{(1)}_{\nu}(z),$$

$$H^{(2)}_{-\nu}(z)=e^{-i\pi\nu}H^{(2)}_{\nu}(z). \tag{12}$$

Finally, we have the inequalities

$$\left|P^{-\mu}_{i\nu-\frac{1}{2}}(\operatorname{ch}\alpha)\right|\leqslant\frac{\Gamma(\frac{1}{2}+\sigma)}{|\Gamma(\frac{1}{2}+\mu)|}P^{-\sigma}_{-\frac{1}{2}}(\operatorname{ch}\alpha),\quad(\alpha\geqslant 0,\ \sigma>-\tfrac{1}{2},\ |\nu|>0),$$

$$\left|Q^{-\mu}_{i\nu-\frac{1}{2}}(\operatorname{ch}\alpha)\right|\leqslant\frac{\Gamma(\frac{1}{2}+\sigma)}{|\Gamma(\frac{1}{2}+\mu)|}Q^{-\sigma}_{-\frac{1}{2}}(\operatorname{ch}\alpha),\quad(\alpha\geqslant 0,\ |\sigma|<\tfrac{1}{2},\ |\nu|>0), \tag{13}$$

where $\mu=\sigma+i\tau$. These follow from (2) and (3).

§2. Inversion of the Integral Equation (1)

We shall now consider equation (1) in detail. First of all, we show that for any function $\varphi_\mu(\nu)\in L(0,\infty)$ the integral (1) defines a continuous function of x in the interval $1<x<\infty$. Indeed, from the analyticity of $P^{-\mu}_{i\nu-\frac{1}{2}}(x)$ we have

$$|\psi_\mu(x+h)-\psi_\mu(x)|\leqslant\int_0^{\infty}\left|P^{-\mu}_{i\nu-\frac{1}{2}}(x+h)-P^{-\mu}_{i\nu-\frac{1}{2}}(x)\right|\cdot|\varphi_\mu(\nu)|\,d\nu<\max_{\nu>0,\ \operatorname{Re}\mu>\frac{1}{2}}\left|P^{-\mu}_{i\nu-\frac{1}{2}}(x+h)-P^{-\mu}_{i\nu-\frac{1}{2}}(x)\right|\int_0^{\infty}|\varphi_\mu(\nu)|\,d\nu<\varepsilon \tag{14}$$

for $h<h_0(\varepsilon)$, $x>1$, $\operatorname{Re}\mu>\frac{1}{2}$;* this proves the continuity of $\psi_\mu(x)$. If the equation (1) is rewritten as

$$\psi_\mu(\operatorname{ch}\alpha)=\int_0^{\infty}P^{-\mu}_{i\nu-\frac{1}{2}}(\operatorname{ch}\alpha)\,\varphi_\mu(\nu)\,d\nu \tag{15}$$

*That the function $|P^{-\mu}_{i\nu-\frac{1}{2}}(x+h)-P^{-\mu}_{i\nu-\frac{1}{2}}(x)|$ is uniformly bounded under the conditions $x\geqslant 1+\delta$, $\nu\geqslant 0$ and $\operatorname{Re}\mu\geqslant\frac{1}{2}+\delta_1$ (δ and $\delta_1>0$ are fixed) can be seen from the integral identity (2) with $\mu=-\mu'$ and $\alpha=\operatorname{arc\,ch}x$.

and use is made of formula (2), then the problem of finding $\varphi_\mu(\nu) \in L(0, \infty)$ for a given continuous function $\psi_\mu(x)$ reduces to solving the Abel equation

$$\sqrt{\frac{2}{\pi}} \frac{\Gamma(\mu+\frac{1}{2})}{sh^{-\mu}\alpha} \psi_\mu(ch\,\alpha) = \int_0^\alpha \frac{\varkappa_\mu(ch\,t)\, sh\,t\, dt}{(ch\,\alpha - ch\,t)^{\frac{1}{2}-\mu}}, \qquad (Re\,\mu > -\tfrac{1}{2}), \tag{16}$$

where

$$\varkappa_\mu(ch\,t)\, sh\,t = \frac{2}{\pi} \int_0^\infty \cos\nu t\, \varphi_\mu(\nu)\, d\nu \tag{17}$$

and then inverting integral (17). We remark that in deriving equation (16) it is necessary to interchange the order of integration with respect to the variables ν and t. This is permissible because of the uniform convergence of the inner integral with respect to the variable ν and the existence [from the condition $\varphi_\mu(\nu) \in L(0, \infty)$] of the outer integral. Under the present hypotheses equation (16) admits a solution in the class of continuous functions and can be written

$$\varkappa_\mu(ch\,t)\, sh\,t = \sqrt{\frac{2}{\pi}} \frac{\cos\mu\pi}{\pi} \Gamma(\mu+\tfrac{1}{2}) \frac{d}{dt} \int_0^t \frac{\psi_\mu(ch\,\alpha)\, sh\,\alpha\, d\alpha}{sh^{-\mu}\alpha\,(ch\,t - ch\,\alpha)^{\mu+\frac{1}{2}}} \qquad (|Re\,\mu| < \tfrac{1}{2}). \tag{18}$$

To find $\varphi_\mu(\nu)$, it remains to apply Fourier inversion to equation (17) in which $\varkappa_\mu(ch\,\alpha)\, sh\,\alpha$ has the value (18). Use is hereby made of the following lemma.

LEMMA.* Let

$$F(\nu) = \int_0^\infty \cos\nu t\, dt \frac{d}{dt} \int_0^t q(ch\,t, ch\,\alpha)\, f(ch\,\alpha)\, sh\,\alpha\, d\alpha$$

be an absolutely convergent iterated integral which for $\nu \geqslant 0$ defines a continuous function $F(\nu)$. Moreover, let $f(ch\,\alpha)(1+ch\,\alpha)^{-\frac{1}{2}} \in L(0, \infty)$, and let the function $q(ch\,t, ch\,\alpha)$ be such that $\varphi(0) = 0$ and $\varphi(\infty) = 0$, where

$$\varphi(t) = \int_0^t q(ch\,t, ch\,\alpha)\, f(ch\,\alpha)\, sh\,\alpha\, d\alpha.$$

Then

$$F(\nu) = \frac{d}{d\nu} \int_0^\infty f(ch\,\alpha)\, sh\,\alpha\, d\alpha \int_0^\nu P(\tau, ch\,\alpha)\, d\tau,$$

*To prove the lemma, it is sufficient to consider the expression $\int_0^\nu F(\nu)\, d\nu$, integrate by parts, and use the Dirichlet rule for changing the order of integration.

where

$$P(\tau, ch\alpha) = \tau \int_{\alpha}^{\infty} q(cht, ch\alpha) \sin\tau t \, dt.$$

As a result, we obtain

$$\varphi_\mu(\nu) = \frac{d}{d\nu} \int_0^\infty \psi_\mu(ch\alpha) sh\alpha \, d\alpha \int_0^\nu \nu th\nu\pi \frac{\Gamma(\frac{1}{2}+\mu+i\nu)\Gamma(\frac{1}{2}+\mu-i\nu)}{\Gamma(\frac{1}{2}+i\nu)\Gamma(\frac{1}{2}-i\nu)} \cdot P^{-\mu}_{i\nu-\frac{1}{2}}(ch\alpha) \, d\alpha \tag{19}$$

$$(|Re\mu| < \tfrac{1}{2}, \quad -\infty < \nu < \infty).$$

If the function $\psi_\mu(x)$ is such that the integral (19) admits differentiation under the integral sign, then

$$\varphi_\mu(\nu) = \nu th\nu\pi \frac{\Gamma(\frac{1}{2}+\mu+i\nu)\Gamma(\frac{1}{2}+\mu-i\nu)}{\Gamma(\frac{1}{2}+i\nu)\Gamma(\frac{1}{2}-i\nu)} \int_0^\infty P^{-\mu}_{i\nu-\frac{1}{2}}(ch\alpha)\psi_\mu(ch\alpha) \cdot sh\alpha \, d\alpha, \tag{20}$$

$$(|Re\mu| < \tfrac{1}{2}, \quad -\infty < \nu < \infty).$$

Formulas (19) and (20) solve the problem of inverting the integral (15) in the strip $|Re\mu| < \frac{1}{2}$. In the case $\mu = 0$, equation (20) gives the well-known result

$$\varphi(\nu) = \nu th\nu\pi \int_0^\infty P_{i\nu-\frac{1}{2}}(ch\alpha)\psi_\mu(ch\alpha) sh\alpha \, d\alpha, \tag{21}$$

$$(-\infty < \nu < \infty),$$

due to Mehler and Fok.

For positive integers $\mu = m$ analogous results can be obtained starting with the classical work of Weyl [4], Titchmarsh [5], and Kodaira [6], and also by generalizing the considerations of Lebedev [7] in the case $\mu = 0$. Taking account of the principle of analytic continuation, we formulate the final result as two theorems.

THEOREM 1. *Let the function $f_\mu(x)$ be such that*

$$|f_\mu(x)| P^{\sigma}_{-\frac{1}{2}}(x) \in L(1, \infty), \quad (-\tfrac{1}{2} < \sigma < \infty),$$

$f_\mu(x)$ is continuous, and has bounded variation in any finite segment of $(1, \infty)$. Then the formula

$$\Phi_\mu(\nu) = \int_1^\infty f_\mu(x) P^{-\mu}_{-\frac{1}{2}+i\nu}(x) \, dx, \tag{22}$$

$$(Re\mu > -\tfrac{1}{2}, \quad -\infty < \nu < \infty)$$

for all $x \geq 1$ implies the inversion formula

$$f_\mu(x) = \int_0^\infty \nu th\nu\pi \frac{\Gamma(\frac{1}{2}+\mu+i\nu)\Gamma(\frac{1}{2}+\mu-i\nu)}{\Gamma(\frac{1}{2}+i\nu)\Gamma(\frac{1}{2}-i\nu)} P^{-\mu}_{-\frac{1}{2}+i\nu}(x)\Phi_\mu(\nu) \, d\nu \tag{23}$$

$$(Re\mu > -\tfrac{1}{2}, \quad 1 \leq x < \infty)$$

For points of discontinuity of first kind the left side of (23) is to be replaced by $\frac{1}{2}[f_\mu(x+0)+f_\mu(x-0)]$.

THEOREM 2. Let $f_\mu(x)$ be any function of $L_2(1,\infty)$, and let $F_\mu(\nu)$ be its integral transform

$$F_\mu(\nu)=\underset{n\to\infty}{l.i.m}\int_1^n f_\mu(x)P_\mu(x,\nu)\,dx,$$

$$P_\mu(x,\nu)=\sqrt{\nu\, th\,\nu\pi\,\frac{\Gamma(\frac{1}{2}+\mu+i\nu)\Gamma(\frac{1}{2}+\mu-i\nu)}{\Gamma(\frac{1}{2}+i\nu)\Gamma(\frac{1}{2}-i\nu)}}\;P^{-\mu}_{i\nu-\frac{1}{2}}(x) \tag{24}$$

$$(Re\,\mu>-\tfrac{1}{2},\quad -\infty<\nu<\infty)$$

Then

$$\int_0^\infty [F_\mu(\nu)]^2\,d\nu=\int_1^\infty f_\mu^2(x)\,dx \tag{25}$$

and almost everywhere

$$f_\mu(x)=\frac{d}{dx}\int_0^\infty F_\mu(\nu)\left\{\int_1^x P_\mu(y,\nu)\,dy\right\}d\nu,$$

$$x\in(1,\infty),\quad Re\,\mu>-\tfrac{1}{2};$$

$$F_\mu(\nu)=\frac{d}{d\nu}\int_1^\infty f_\mu(x)\left\{\int_0^\nu P_\mu(x,\nu)\,d\nu\right\}dx, \tag{26}$$

$$\nu\in(0,\infty),\quad Re\,\mu>-\tfrac{1}{2};$$

In conclusion, we consider an example. We put

$$\psi_\mu(ch\,\alpha)=sh^\mu\alpha\, e^{-a\,ch\,\alpha},\quad a>0,\quad Re\,\mu>-\tfrac{1}{2}.$$

Here

$$\varphi_\mu(\nu)=\nu\,th\,\nu\pi\,\frac{\Gamma(\frac{1}{2}+\mu+i\nu)\Gamma(\frac{1}{2}+\mu-i\nu)}{\Gamma(\frac{1}{2}+i\nu)\Gamma(\frac{1}{2}-i\nu)}\sqrt{\frac{2}{\pi a}}\,\frac{K_{i\nu}(a)}{a^\mu}$$

and hence

$$e^{-a\,ch\,\alpha}sh^\mu\alpha=\int_0^\infty \nu\,th\,\nu\pi\,\frac{\Gamma(\frac{1}{2}+\mu+i\nu)\Gamma(\frac{1}{2}+\mu-i\nu)}{\Gamma(\frac{1}{2}+i\nu)\Gamma(\frac{1}{2}-i\nu)}\sqrt{\frac{2}{\pi a}}\,\frac{K_{i\nu}(a)}{a^\mu}\cdot P^{-\mu}_{i\nu-\frac{1}{2}}(ch\,\alpha)\,d\nu,$$

$$\sqrt{\frac{2}{\pi a}}\,\frac{K_{i\nu}(a)}{a^\mu}=\int_0^\infty sh^{\mu+1}\alpha\, e^{-a\,ch\,\alpha}\,d\alpha,$$

which can also easily be verified directly.

The author wishes to thank V. M. Babich for reading through this work and making a number of valuable suggestions.

LITERATURE CITED

1. Mehler, F. G., Math. Ann., Vol. 18 (1881).
2. Fok, V. A., Dokl. Akad. Nauk SSSR, Vol. 39, No. 7 (1943).
3. Hobson, E. W., The Theory of Spherical and Ellipsoidal Harmonics, Cambridge Univ. Press, Cambridge (1931)
4. Weyl, H., J. rein. angew. Math., Vol. 141 (1912).
5. Titchmarsh, E. C., Eigenfunction Expansions Associated with Second-Order Differential Equations, Clarendon Press, Oxford (1950).
6. Kodaira, K., Am. J. Math., Vol. 72 (1950).
7. Lebedev, N. N., Certain Integral Transforms in Mathematical Physics, Dissertation [in Russian], (1951).

APPLICATION OF AN INTEGRAL TRANSFORM WITH GENERALIZED LEGENDRE KERNEL TO THE SOLUTION OF INTEGRAL EQUATIONS WITH SYMMETRIC KERNELS

B. G. Nikolaev

Besides application to integration of partial differential equations, the method of integral transforms can be used to obtain closed forms for the solutions of linear integral equations with symmetric kernels.

In this paper we restrict our consideration to two integral equations whose solutions can be written down using the integral transform studied in [1]. As a first example, we consider the equation

$$\varphi_\mu(x) = g_\mu(x) + \lambda \int_1^\infty \frac{(x^2-1)^{\frac{\mu}{2}}(y^2-1)^{\frac{\mu}{2}}}{(x+y)^{1+2\mu}} \varphi_\mu(y)\,dy \tag{1}$$

$$\left(1 \leqslant x < \infty, \quad |\lambda| < \frac{\Gamma(2\mu+1)}{\Gamma^2(\mu+\frac{1}{2})}, \quad \operatorname{Re}\mu > -\frac{1}{2}\right).$$

A formal solution of equation (1) is obtained as follows. We assume that there exist functions

$$\Phi_\mu(\nu) = \int_1^\infty \varphi_\mu(x) P^{-\mu}_{-\frac{1}{2}+i\nu}(x)\,dx,$$

$$G_\mu(\nu) = \int_1^\infty g_\mu(x) P^{-\mu}_{-\frac{1}{2}+i\nu}(x)\,dx. \tag{2}$$

We multiply both sides of equation (1) by $P^{-\mu}_{-\frac{1}{2}+i\nu}(x)$ and integrate the result so obtained with respect to x over the integral from 1 to ∞. Using the identity*

---

*This identity can be easily verified by using the well-known representation

$$P^{-\mu}_{-\frac{1}{2}+i\nu}(\operatorname{ch}\alpha) = \sqrt{\frac{2}{\pi}} \frac{\Gamma(\mu+\frac{1}{2})\operatorname{sh}^\mu\alpha}{\Gamma(\frac{1}{2}+\mu+i\nu)\Gamma(\frac{1}{2}+\mu-i\nu)} \int_0^\infty \frac{\cos\nu\psi\,d\psi}{(\operatorname{ch}\alpha+\operatorname{ch}\psi)^{\mu+\frac{1}{2}}}$$

for the associated Legendre function.

$$P^{-\mu}_{-\frac{1}{2}+i\nu}(y)=\frac{\Gamma(1+2\mu)}{\Gamma(\frac{1}{2}+\mu+i\nu)\,\Gamma(\frac{1}{2}+\mu-i\nu)}\int_{1}^{\infty}\frac{(x^2-1)^{\frac{\mu}{2}}(y^2-1)^{\frac{\mu}{2}}}{(x+y)^{1+2\mu}}\cdot P^{-\mu}_{-\frac{1}{2}+i\nu}(x)\,dx, \tag{3}$$

$$(\operatorname{Re}\mu>-\tfrac{1}{2},\quad y\geq 0),$$

we obtain

$$\Phi_\mu(\nu)=G_\mu(\nu)+\frac{\lambda\,\Gamma(\frac{1}{2}+\mu+i\nu)\,\Gamma(\frac{1}{2}+\mu-i\nu)\,\Phi_\mu(\nu)}{\Gamma(1+2\mu)}. \tag{4}$$

From this it follows that

$$\Phi_\mu(\nu)=\frac{G_\mu}{1-\dfrac{\lambda\,\Gamma(\frac{1}{2}+\mu+i\nu)\,\Gamma(\frac{1}{2}+\mu-i\nu)}{\Gamma(1+2\mu)}}. \tag{5}$$

The desired solution can be obtained by the inversion formula [1]:

$$\varphi_\mu(x)=\int_{0}^{\infty}\frac{\nu\,\mathrm{th}\,\nu\pi\,\dfrac{\Gamma(\frac{1}{2}+\mu+i\nu)\,\Gamma(\frac{1}{2}+\mu-i\nu)}{\Gamma(\frac{1}{2}+i\nu)\,\Gamma(\frac{1}{2}-i\nu)}\,G_\mu(\nu)\,P^{-\mu}_{-\frac{1}{2}+i\nu}(x)}{1-\dfrac{\lambda\,\Gamma(\frac{1}{2}+\mu+i\nu)\,\Gamma(\frac{1}{2}+\mu-i\nu)}{\Gamma(1+2\mu)}}\cdot d\nu. \tag{6}$$

It remains to formulate conditions under which formula (6) actually defines a solution of equation (1).

THEOREM 1. *Let $g_\mu(x)$ be a given continuous function which has bounded variation in any finite segment of $(1,\infty)$ and which satisfies the following conditions:*

$$(*)\qquad |g_\mu(x)|\,P^{-\sigma}_{-\frac{1}{2}}(x)\in L(1,\infty),\quad(-\tfrac{1}{2}<\sigma<\infty),$$

$$(**)\qquad \nu\,|G_\mu(\nu)|\in L(0,\infty).$$

Then the integral (6) exists and defines a continuous solution of the integral equation (1).

We note first of all that the estimate

$$\left|\frac{\nu\,\mathrm{th}\,\nu\pi\,\dfrac{\Gamma(\frac{1}{2}+\mu+i\nu)\,\Gamma(\frac{1}{2}+\mu-i\nu)}{\Gamma(\frac{1}{2}+i\nu)\,\Gamma(\frac{1}{2}-i\nu)}\,G_\mu(\nu)\,P^{-\mu}_{-\frac{1}{2}+i\nu}(x)}{1-\dfrac{\Gamma(\frac{1}{2}+\mu+i\nu)\,\Gamma(\frac{1}{2}+\mu-i\nu)}{\Gamma(1+2\mu)}}\right|<\varkappa_\mu(\lambda)\,P^{-\sigma}_{-\frac{1}{2}}(x)\,\nu\,|G_\mu(\nu)|,\quad(-\tfrac{1}{2}<\sigma<\infty), \tag{7}$$

where†

$$\varkappa_\mu(\lambda)=\frac{C_\mu}{1-|\lambda|\dfrac{\Gamma^2(\mu+\frac{1}{2})}{\Gamma(1+2\mu)}},\qquad |\lambda|<\frac{\Gamma(1+2\mu)}{\Gamma^2(\frac{1}{2}+\mu)} \tag{8}$$

†We note that the identity $\Gamma(\frac{1}{2}+\mu+i\nu)\,\Gamma(\frac{1}{2}+\mu-i\nu)=\frac{\Gamma(1+2\mu)}{2^{1+2\mu}}\int_0^\infty\frac{\cos\nu x\,dx}{\mathrm{ch}^{1+2\mu}x}$ implies $|\Gamma(\frac{1}{2}+\mu+i\nu)\,\Gamma(\frac{1}{2}+\mu-i\nu)|\leq\Gamma(\mu+\frac{1}{2})$. Therefore, (3) is solvable.

and condition (**) imply the existence and continuity of the function $\varphi_\mu(x)$, defined by equation (6).

Further, we find

$$\lambda\int_1^\infty \frac{(x^2-1)^{\frac{\mu}{2}}(y^2-1)^{\frac{\mu}{2}}}{(x+y)^{1+2\mu}}\varphi_\mu(y)\,dy=\lambda\int_0^\infty \frac{\nu\,\mathrm{th}\,\nu\pi\,\dfrac{\Gamma(\frac12+\mu+i\nu)\Gamma(\frac12+\mu-i\nu)}{\Gamma(\frac12+i\tau)\Gamma(\frac12-i\tau)}\cdot G_\mu(\nu)}{1-\dfrac{\lambda\Gamma(\frac12+\mu+i\tau)\Gamma(\frac12+\mu-i\nu)}{\Gamma(1+2\mu)}}\,d\nu\cdot$$

$$\int_1^\infty \frac{(x^2-1)^{\frac{\mu}{2}}(y^2-1)^{\frac{\mu}{2}}}{(x+y)^{1+2\mu}}\cdot P^{-\mu}_{-\frac12+i\nu}(y)\,dy=-\int_0^\infty \nu\,\mathrm{th}\,\nu\pi\,\frac{\Gamma(\frac12+\mu+i\nu)\Gamma(\frac12+\mu-i\nu)}{\Gamma(\frac12+i\nu)\Gamma(\frac12-i\nu)}G_\mu(\nu)P^{-\mu}_{-\frac12+i\nu}(x)\,d\nu+ \tag{9}$$

$$+\int_0^\infty \frac{\nu\,\mathrm{th}\,\nu\pi\,\dfrac{\Gamma(\frac12+\mu+i\nu)\Gamma(\frac12+\mu-i\nu)}{\Gamma(\frac12+i\tau)\Gamma(\frac12-i\nu)}G_\mu(\nu)}{1-\dfrac{\lambda\Gamma(\frac12+\mu+i\nu)\Gamma(\frac12+\mu-i\nu)}{\Gamma(1+2\mu)}}P^{-\mu}_{-\frac12+i\nu}(x)\,d\nu=-g_\mu(x)+\varphi_\mu(x)\,.$$

The justification for changing the order of the integrals of (9) follows from the conditions imposed on the function $g_\mu(x)$.

As another example, we consider the equation

$$\varphi_\mu(x)=g_\mu(x)+\lambda\int_1^\infty \frac{(x^2-1)^{\frac{\mu}{2}}(y^2-1)^{\frac{\mu}{2}}P^{-\mu}_{-\frac12+i\nu}(y)}{(x^2+y^2-1)^{\mu+\frac12}}\varphi_\mu(y)\,dy,$$

$$\left(1\le x<\infty,\quad |\lambda|<\frac{2\Gamma(\mu+\frac12)}{\Gamma^2(\frac{\mu}{2}+\frac14)},\quad \mathrm{Re}\,\mu>-\tfrac12\right). \tag{10}$$

Proceeding in a similar manner, we obtain

$$\Phi_\mu(\nu)=G_\mu(\nu)+\frac{\lambda\Gamma(\frac{\mu}{2}+\frac14+\frac{i\nu}{2})\Gamma(\frac{\mu}{2}+\frac14-\frac{i\nu}{2})}{2\Gamma(\mu+\frac12)}\Phi_\mu(\nu) \tag{11}$$

or

$$\varphi_\mu(x)=\int_0^\infty \frac{\nu\,\mathrm{th}\,\nu\pi\,\dfrac{\Gamma(\frac12+\mu+i\nu)\Gamma(\frac12+\mu-i\nu)}{\Gamma(\frac12+i\nu)\Gamma(\frac12-i\nu)}G_\mu(\nu)P^{-\mu}_{-\frac12+i\nu}(x)}{1-\dfrac{\lambda\Gamma(\frac{\mu}{2}+\frac14+\frac{i\nu}{2})\Gamma(\frac{\mu}{2}+\frac14-\frac{i\nu}{2})}{2\Gamma(\mu+\frac12)}}\cdot d\nu\,. \tag{12}$$

Just as in the preceding case, it is easy to verify this formula if the function $g_\mu(x)$ satisfies the conditions of Theorem 1 and use is made of the identity

$$P^{-\mu}_{-\frac12+i\nu}(y)=\frac{2\Gamma(\mu+\frac12)}{\Gamma(\frac{\mu}{2}+\frac14+\frac{i\nu}{2})\Gamma(\frac{\mu}{2}+\frac14-\frac{i\nu}{2})}\int_0^\infty \frac{(x^2-1)^{\frac{\mu}{2}}(y^2-1)^{\frac{\mu}{2}}P^{-\mu}_{-\frac12+i\nu}(x)\,dx}{(x^2+y^2-1)^{\mu+\frac12}} \tag{13}$$

$$(\mathrm{Re}\,\mu>-\tfrac12,\quad y\ge 0)\,.$$

We remark that in the particular case $\mu=0$ equations (1) and (10) have been studied in [2].

The author wishes to thank V. M. Babich for reading through this work and making a number of valuable suggestions.

LITERATURE CITED

1. Nikolaev, B. G., "The expansion of an arbitrary function in terms of an integral of associated Legendre functions of first kind with complex index," this volume, p. 45.
2. Lebedev, N. N., Certain Integral Transforms in Mathematical Physics, Dissertation [in Russian] (1951).

SOLUTION OF THREE-DIMENSIONAL PROBLEMS FOR THE HYPERBOLOID OF REVOLUTION AND THE LENS IN ELECTRICAL PROSPECTING

B. G. Nikolaev

In this paper solutions are given for problems of the potential distribution created by a point source of current in the case of a hyperbolic or spherical boundary of separation. The solutions are obtained by integrating the Laplace equation in degenerate ellipsoidal (toroidal) coordinates. To find the coefficients, a generalization of the Mehler-Fok integral theorem to the case $m \neq 0$ is used. The final expressions for the potential functions are given in the form of series whose coefficients are given by real integrals.

The Field of a Point Source Located on the Boundary of a Half Space Above a Conducting Hyperbolic Dome

We suppose that in a homogeneous, isotropic space with conductivity σ_1 there is an inclusion in the form of a two-sheeted hyperboloid of revolution with conductivity σ_2. We place a point source J in its plane of symmetry and shall find the resulting potential distribution.

We introduce a system of degenerate ellipsoidal coordinates α, β, φ, related to spherical coordinates r, z, φ by the equations (Fig. 1):

$$\left.\begin{aligned} r &= c\,\operatorname{sh}\alpha\,\sin\beta \\ z &= c\,\operatorname{ch}\alpha\,\cos\beta \\ \varphi &= \varphi \end{aligned}\right\} \tag{1}$$

$$(0 \leqslant \alpha, \quad 0 \leqslant \beta \leqslant \pi, \quad -\pi < \varphi \leqslant \pi).$$

Then the region $0 \leqslant \alpha < \infty$, $\beta_0 < \beta < \pi - \beta_0$, $-\pi < \varphi \leqslant \pi$ is characterized by conductivity σ_1; the space occupied by the inclusion $0 \leqslant \alpha < \infty$, $0 \leqslant \beta \leqslant \beta_0$, $-\pi < \varphi \leqslant \pi$ and $0 \leqslant \alpha < \infty$, $\pi - \beta_0 \leqslant \beta \leqslant \pi$, $-\pi < \varphi \leqslant \pi$ has conductivity σ_2.

We denote by $U_0(\alpha, \beta, \varphi)$ the potential function of the primary field created by the point source in an unbounded medium with conductivity σ_1, and by

$$\begin{aligned} &U_1(\alpha,\beta,\varphi), \quad \text{for} \quad (\alpha \geqslant 0, \quad \beta_0 < \beta < \pi - \beta_0, \quad -\pi < \varphi \leqslant \pi) \\ &U_2(\alpha,\beta,\varphi), \quad \text{for} \quad (\alpha \geqslant 0, \quad 0 \leqslant \beta \leqslant \beta_0, \quad -\pi < \varphi \leqslant \pi) \\ &U_3(\alpha,\beta,\varphi), \quad \text{for} \quad (\alpha \geqslant 0, \quad \pi - \beta_0 \leqslant \beta \leqslant \pi, \quad -\pi < \varphi \leqslant \pi) \end{aligned} \tag{2}$$

the potential functions of the secondary field.

The problem of determining the quantities $U_n(\alpha, \beta, \varphi)$, $n = 1, 2, 3$, reduces to solution of the Laplace equation

$$\frac{1}{\operatorname{sh}\alpha}\frac{\partial}{\partial\alpha}\left(\operatorname{sh}\alpha\frac{\partial U_n}{\partial\alpha}\right) + \frac{1}{\sin\beta}\frac{\partial}{\partial\beta}\left(\sin\beta\frac{\partial U_n}{\partial\beta}\right) + \left(\frac{1}{\operatorname{sh}^2\alpha} + \frac{1}{\sin^2\beta}\right)\frac{\partial^2 U_n}{\partial\varphi^2} = 0, \quad (n = 1, 2, 3) \tag{3}$$

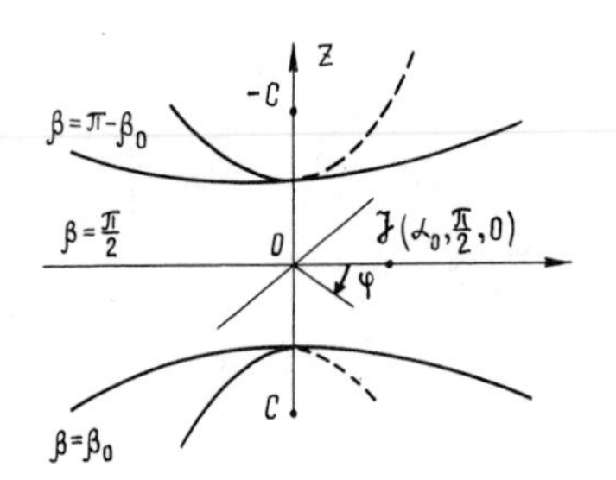

Fig. 1

with boundary conditions

$$(U_0 + U_1)_{\beta=\beta_0} = (U_2)_{\beta=\beta_0},$$
$$\sigma_1 \frac{\partial}{\partial\beta}(U_0+U_1)_{\beta=\beta_0} = \sigma_2\left(\frac{\partial U_2}{\partial\beta}\right)_{\beta=\beta_0} \tag{4}$$

and conditions on the decrease of the functions $U_n(\alpha,\beta,\varphi)$ at infinite.

Here the field $U_0(\alpha,\beta,\varphi)$ is determined by the point source and has the form

$$U_0 = \frac{J}{4\pi\sigma_1 c\sqrt{ch^2\alpha - \sin^2\beta + sh^2\alpha_0 - 2\,sh\alpha\, sh\alpha_0 \cos\varphi}} = \sum_{m=0}^{\infty} \varepsilon_m \cos m\varphi \frac{J\, Q_{m-\frac{1}{2}}\left(\frac{ch^2\alpha - \sin^2\beta + sh^2\alpha_0}{2\,sh\alpha\, sh\alpha_0 \sin\beta}\right)}{4\pi^2\sigma_1 c\sqrt{sh\alpha\, sh\alpha_0 \sin\beta}}, \tag{5}$$

$$(\varepsilon_0 = \tfrac{1}{2},\quad \varepsilon_m = 1,\quad m = 1, 2, \ldots),$$

where J is the magnitude of the current and $Q_p(z)$ is a Legendre function of second kind.

The method of separation of variables leads to the following system of particular solutions for equation (3) which depend on the parameters ν and m:

$$u = [M_m(\nu) P^{-m}_{-\frac{1}{2}+i\nu}(ch\alpha) + N_m(\nu) P^{-m}_{-\frac{1}{2}+i\nu}(-ch\alpha)] \cdot [A_m(\nu) P^{-m}_{-\frac{1}{2}+i\nu}(\cos\beta) + C_m(\nu) P^{-m}_{-\frac{1}{2}+i\nu}(-\cos\beta)] \cos m\varphi, \tag{6}$$

where $P^q_p(z)$ are associated Legendre functions of first kind.

We shall seek the solution of problem (3)-(5) in the form [1]:

$$U_1 = \sum_{m=0}^{\infty} \varepsilon_m \cos m\varphi \int_0^{\infty} [A_m(\nu) P^{-m}_{-\frac{1}{2}+i\nu}(\cos\beta) + C_m(\nu) P^{-m}_{-\frac{1}{2}+i\nu}(-\cos\beta)] \cdot P^{-m}_{-\frac{1}{2}+i\nu}(ch\alpha)\, d\nu,$$

$$U_2 = \sum_{m=0}^{\infty} \varepsilon_m \cos m\varphi \int_0^{\infty} B_m(\nu) P^{-m}_{-\frac{1}{2}+i\nu}(\cos\beta) P^{-m}_{-\frac{1}{2}+i\nu}(ch\alpha)\, d\nu, \tag{7}$$

$$U_3 = \sum_{m=0}^{\infty} \varepsilon_m \cos m\varphi \int_0^{\infty} D_m(\nu) P^{-m}_{-\frac{1}{2}+i\nu}(-\cos\beta) P^{-m}_{-\frac{1}{2}+i\nu}(ch\alpha)\, d\nu,$$

$$(\varepsilon_0 = \tfrac{1}{2},\quad \varepsilon_m = 1,\quad m = 1, 2, \ldots).$$

By the symmetry of the function $U_1(\alpha,\beta,\varphi)$ relative to the plane $\beta = \frac{\pi}{2}$ we have:

$$\left.\frac{\partial U_1}{\partial\beta}\right|_{\beta=\frac{\pi}{2}} = 0. \tag{8}$$

The last condition with the equation

$$P^{\kappa\prime}_p(0) = \frac{2^{\kappa+1}}{\sqrt{\pi}} \frac{\Gamma\left(\frac{p+\kappa}{2}+1\right)}{\Gamma\left(\frac{p-\kappa+1}{2}\right)} \sin\frac{(p-\kappa)\pi}{2} \neq 0 \tag{9}$$

valid for any $\frac{p-\kappa}{2} \neq n$, where $n = 0, \pm 1, \pm 2, \cdots$, requires that $A_m(\nu) = c_m(\nu)$.

On the other hand, the identity

$$P^{-m}_{-\frac{1}{2}+i\nu}[-\cos(\pi-\beta_0)] = P^{-m}_{-\frac{1}{2}+i\nu}(\cos\beta_0) \tag{10}$$

leads to the condition $B_m(\nu) = \mathcal{D}_m(\nu)$.

As a result, we seek a solution in the form:

$$u_1 = \sum_{m=0}^{\infty} \varepsilon_m \cos m\varphi \int_0^\infty A_m(\nu)\left[P^{-m}_{-\frac{1}{2}+i\nu}(\cos\beta) + P^{-m}_{-\frac{1}{2}+i\nu}(-\cos\beta)\right] \cdot P^{-m}_{-\frac{1}{2}+i\nu}(ch\,\alpha)\,d\nu,$$

$$u_2 = \sum_{m=0}^{\infty} \varepsilon_m \cos m\varphi \int_0^\infty B_m(\nu) P^{-m}_{-\frac{1}{2}+i\nu}(\cos\beta) P^{-m}_{-\frac{1}{2}+i\nu}(ch\,\alpha)\,d\nu, \tag{11}$$

$$(\varepsilon_0 = \tfrac{1}{2}, \quad \varepsilon_m = 1, \quad m = 1, 2, \ldots).$$

Substituting (11) into (4) and using (5), we obtain

$$\frac{\mathcal{J}\, Q_{m-\frac{1}{2}}(\ldots)\big|_{\beta=\beta_0}}{4\pi^2\sigma_1 c\sqrt{sh\,\alpha\, sh\,\alpha_0 \sin\beta_0}} + \int_0^\infty A_m(\nu)\left[P^{-m}_{-\frac{1}{2}+i\nu}(\cos\beta_0) + P^{-m}_{-\frac{1}{2}+i\nu}(-\cos\beta_0)\right] P^{-m}_{-\frac{1}{2}+i\nu}(ch\,\alpha)\,d\nu = \int_0^\infty B_m(\nu) P^{-m}_{-\frac{1}{2}+i\nu}(\cos\beta_0) P^{-m}_{-\frac{1}{2}+i\nu}(ch\,\alpha)\,d\nu,$$

$$-\mathcal{J}\frac{d}{d\beta_0}\frac{Q_{m-\frac{1}{2}}(\ldots)\big|_{\beta=\beta_0}}{\sqrt{sh\,\alpha\, sh\,\alpha_0 \sin\beta_0}} + \sigma_1\int_0^\infty A_m(\nu)\left[P^{-m'}_{-\frac{1}{2}+i\nu}(-\cos\beta_0) - P^{-m'}_{-\frac{1}{2}+i\nu}(\cos\beta_0)\right] P^{-m}_{-\frac{1}{2}+i\nu}(ch\,\alpha)\,d\nu = \sigma_2\int_0^\infty B_m(\nu) P^{-m'}_{-\frac{1}{2}+i\nu}(\cos\beta_0) P^{-m}_{-\frac{1}{2}+i\nu}(ch\,\alpha)\,d\nu. \tag{12}$$

To determine the coefficients, we use the integral theorem for expansion in terms of associated Legendre functions of first kind [2]:

$$f_m(\nu) = \nu\, th\,\nu\pi\, \frac{\Gamma(\frac{1}{2}+m+i\nu)\Gamma(\frac{1}{2}+m-i\nu)}{\Gamma(\frac{1}{2}+i\nu)\Gamma(\frac{1}{2}-i\nu)} \cdot \int_0^\infty P^{-m}_{-\frac{1}{2}+i\nu}(ch\,\alpha)\, sh\,\alpha\, d\alpha \int_0^\infty f_m(\tau) P^{-m}_{-\frac{1}{2}+i\tau}(ch\,\alpha)\,d\tau. \tag{13}$$

Multiplying both sides of (12) by $\nu\, th\,\nu\pi \frac{\Gamma(\frac{1}{2}+m+i\nu)\Gamma(\frac{1}{2}+m-i\nu)}{\Gamma(\frac{1}{2}+i\nu)\Gamma(\frac{1}{2}-i\nu)} \cdot P^{-m}_{-\frac{1}{2}+i\nu}(ch\,\alpha)\, sh\,\alpha$ and integrating with respect to α between the limits 0 and ∞ we obtain the following system:

$$A_m(\nu)\left[P^{-m}_{-\frac{1}{2}+i\nu}(\cos\beta_0) + P^{-m}_{-\frac{1}{2}+i\nu}(-\cos\beta_0)\right] - B_m(\nu) P^{-m}_{-\frac{1}{2}+i\nu}(\cos\beta_0) = I^{(1)}_m,$$

$$\sigma_1 A_m(\nu)\left[-P^{-m'}_{-\frac{1}{2}+i\nu}(\cos\beta_0) + P^{-m'}_{-\frac{1}{2}+i\nu}(-\cos\beta_0)\right] + \sigma_2 B_m(\nu) P^{-m'}_{-\frac{1}{2}+i\nu}(\cos\beta_0) = I^{(2)}_m, \tag{14}$$

where

$$I^{(1)}_m = \frac{\nu\, th\,\nu\pi\, \Gamma(\frac{1}{2}+m+i\nu)\Gamma(\frac{1}{2}+m-i\nu)\mathcal{J}}{4\pi^2\sigma_1 c\, \Gamma(\frac{1}{2}+i\nu)\Gamma(\frac{1}{2}-i\nu)\sqrt{sh\,\alpha_0}} \cdot \int_0^\infty \frac{Q_{m-\frac{1}{2}}(\ldots)\big|_{\beta=\beta_0}}{\sqrt{sh\,\alpha \sin\beta_0}} P^{-m}_{-\frac{1}{2}+i\nu}(ch\,\alpha)\, sh\,\alpha\, d\alpha,$$

$$I^{(2)}_m = -\frac{\nu\, th\,\nu\pi\, \Gamma(\frac{1}{2}+m+i\nu)\Gamma(\frac{1}{2}+m-i\nu)\mathcal{J}}{4\pi^2 c\, \Gamma(\frac{1}{2}+i\nu)\Gamma(\frac{1}{2}-i\nu)\sin\beta_0\sqrt{sh\,\alpha_0}} \cdot \int_0^\infty \frac{d}{d\beta_0}\frac{Q_{m-\frac{1}{2}}(\ldots)\big|_{\beta=\beta_0}}{\sqrt{sh\,\alpha \sin\beta_0}} \cdot P^{-m}_{-\frac{1}{2}+i\nu}(ch\,\alpha)\, sh\,\alpha\, d\alpha = -\frac{\sigma_1}{\sin\beta_0}\frac{d}{d\beta_0} I^{(1)}_m. \tag{15}$$

The desired coefficient is given by

$$A_m(\nu) = \frac{\sigma_2 I_m^{(1)} P^{-m'}_{-\frac{1}{2}+i\nu}(\cos\beta_0) + I_m^{(2)} P^{-m}_{-\frac{1}{2}+i\nu}(\cos\beta_0)}{\sigma_2 \Delta_m^{(1)} P^{-m'}_{-\frac{1}{2}+i\nu}(\cos\beta_0) + \sigma_1 \Delta_m^{(2)} P^{-m}_{-\frac{1}{2}+i\nu}(\cos\beta_0)} \tag{16}$$

where

$$\Delta_m^{(1)} = P^{-m}_{-\frac{1}{2}+i\nu}(\cos\beta_0) + P^{-m}_{-\frac{1}{2}+i\nu}(-\cos\beta_0),$$

$$\Delta_m^{(2)} = P^{-m'}_{-\frac{1}{2}+i\nu}(\cos\beta_0) - P^{-m'}_{-\frac{1}{2}+i\nu}(-\cos\beta_0). \tag{16$_1$}$$

The solution for the function in which we are interested is represented in the form

$$u_1 = \sum_{m=0}^{\infty} \varepsilon_m \cos m\varphi \int_0^{\infty} A_m(\nu) P^{-m}_{-\frac{1}{2}+i\nu}(\operatorname{ch}\alpha)\left[P^{-m}_{-\frac{1}{2}+i\nu}(\cos\beta) + P^{-m}_{-\frac{1}{2}+i\nu}(-\cos\beta)\right] d\nu, \tag{17}$$

$$\left(\varepsilon_0 = \frac{1}{2},\quad \varepsilon_m = 1,\quad m = 1, 2, \ldots\right).$$

The value of the potential is expressed by the sum

$$u = \frac{J}{4\pi\sigma_1 c\sqrt{\operatorname{ch}^2\alpha - \sin^2\beta + \operatorname{sh}^2\alpha_0 - 2\operatorname{sh}\alpha\operatorname{sh}\alpha_0 \sin\beta\cos\varphi}} + \sum_{m=0}^{\infty}\varepsilon_m \cos m\varphi \int_0^{\infty} A_m(\nu) P^{-m}_{-\frac{1}{2}+i\nu}(\operatorname{ch}\alpha)\left[P^{-m}_{-\frac{1}{2}+i\nu}(\cos\beta) +\right.$$

$$\left. + P^{-m}_{-\frac{1}{2}+i\nu}(-\cos\beta)\right] d\nu, \quad \left(\varepsilon_0 = \frac{1}{2},\quad \varepsilon_m = 1,\quad m = 1, 2, \ldots\right). \tag{18}$$

We shall express the variables α and β in (18) in terms of the variables r and z of the cylindrical coordinate system. From equation (1) we have

$$r = c\operatorname{sh}\alpha \sin\beta,$$

$$z = c\operatorname{ch}\alpha\cos\beta, \tag{19}$$

whence for $r = 0$, $|z| \leqslant c$

$$\alpha = 0, \quad z = h = c\cos\beta \tag{20}$$

and hence

$$\cos\beta_0 = \frac{h}{c} \tag{21}$$

(h is the distance from the boundary of the half space to the top of the dome).

On the boundary surface $r\big|_{\beta=\frac{\pi}{2}} = c\operatorname{sh}\alpha$, such that $\operatorname{ch}\alpha = \sqrt{1 + \frac{r^2}{c^2}}$ and (18) takes the form:

$$u = \frac{J}{2\pi\sigma_1\sqrt{r^2 + c^2\operatorname{sh}^2\alpha_0 - 2cr\operatorname{sh}\alpha_0\cos\varphi}} + 4\sum_{m=0}^{\infty}\varepsilon_m\cos m\varphi \cdot \int_0^{\infty} \frac{\left[\sigma_2 I_m^{(1)} P^{-m'}_{-\frac{1}{2}+i\nu}\left(\frac{h}{c}\right) + I_m^{(2)} P^{-m}_{-\frac{1}{2}+i\nu}\left(\frac{h}{c}\right)\right] P^{-m}_{-\frac{1}{2}+i\nu}\left(\sqrt{1+\frac{r^2}{c^2}}\right) d\nu}{\sigma_2 \Delta_m^{(1)} P^{-m'}_{-\frac{1}{2}+i\nu}\left(\frac{h}{c}\right) + \sigma_1 \Delta_m^{(2)} P^{-m}_{-\frac{1}{2}+i\nu}\left(\frac{h}{c}\right)}, \tag{22}$$

$$\Delta_m^{(1)} = P^{-m}_{-\frac{1}{2}+i\nu}\left(\frac{h}{c}\right) + P^{-m}_{-\frac{1}{2}+i\nu}\left(-\frac{h}{c}\right), \quad \Delta_m^{(2)} = P^{-m'}_{-\frac{1}{2}+i\nu}\left(\frac{h}{c}\right) - P^{-m'}_{-\frac{1}{2}+i\nu}\left(-\frac{h}{c}\right),$$

$$\left(\varepsilon_0 = \frac{1}{2},\quad \varepsilon_m = 1,\quad m = 0, 1, 2, \ldots\right).$$

We remark that a similar problem for the case of rotational symmetry when the desired function u is independent of φ was considered in [3].

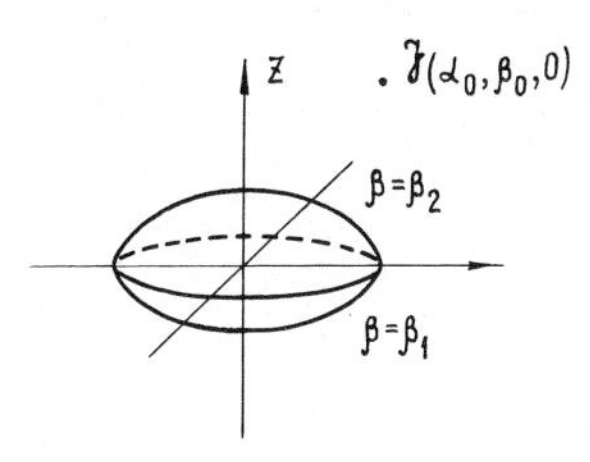

Fig. 2

The Field of a Point Source Located at an Arbitrary Point of Space Above an Ideally Conducting Lens

We consider an inclusion in the form of an ideally conducting lens in a homogeneous isotropic space with conductivity σ. The point source from which there flows a current $\mathcal{J}$ is located at an arbitrary point of the space with conductivity σ. We seek the potential distribution outside the lens.

We introduce a toroidal coordinate system α, β, φ, related to the cylindrical coordinates r, z, φ by the equations (Fig. 2):

$$r=\frac{c\,\mathrm{sh}\,\alpha}{\mathrm{ch}\,\alpha-\cos\beta},\quad z=\frac{c\sin\beta}{\mathrm{ch}\,\alpha-\cos\beta},\quad \varphi=\varphi,\quad (\alpha\geqslant 0), \tag{23}$$

$$\beta_0<\beta\leqslant 2\pi+\beta_0,\ -\pi<\varphi\leqslant\pi,\ \beta_0\ -$$

Then the region $0\leqslant\alpha<\infty$, $\beta_2\leqslant\beta\leqslant 2\pi+\beta_1$, $-\pi<\varphi\leqslant\pi$ is characterized by conductivity σ; the part of the space $0\leqslant\alpha<\infty$, $\beta_1<\beta<\beta_2$, $-\pi<\varphi\leqslant\pi$ has infinite conductivity.

We denote by $u_1(\alpha,\beta,\varphi)$, where $\alpha\geqslant 0$, $\beta_2\leqslant\beta\leqslant 2\pi+\beta_1$, $-\pi<\varphi\leqslant\pi$, the potential function of the secondary field. This function is a solution of the Laplace equation:

$$\frac{\partial}{\partial\alpha}\left(\frac{\mathrm{sh}\,\alpha}{\mathrm{ch}\,\alpha-\cos\beta}\frac{\partial u_1}{\partial\alpha}\right)+\frac{\partial}{\partial\beta}\left(\frac{\mathrm{sh}\,\alpha}{\mathrm{ch}\,\alpha-\cos\beta}\frac{\partial u_1}{\partial\beta}\right)+\frac{1}{(\mathrm{ch}\,\alpha-\cos\beta)\,\mathrm{sh}\,\alpha}\frac{\partial^2 u_1}{\partial\varphi^2}=0 \tag{24}$$

with the following boundary conditions

$$(u_0+u_1)_{\beta=\beta_2}=0,\quad (u_0+u_1)_{\beta=2\pi+\beta_1}=0, \tag{25}$$

where

$$u_0=\frac{\mathcal{J}}{4\pi\sigma\sqrt{r^2+r_0^2-2rr_0\cos\varphi+(z-z_0)^2}}=\sum_{m=0}^{\infty}\varepsilon_m\cos m\varphi\,\frac{\mathcal{J}\,Q_{m-\frac{1}{2}}\left[\frac{r^2+r_0^2+(z-z_0)^2}{2rr_0}\right]}{4\pi^2\sigma\sqrt{rr_0}}, \tag{26}$$

$$r=\frac{c\,\mathrm{sh}\,\alpha}{\mathrm{ch}\,\alpha-\cos\beta},\quad r_0=\frac{c\,\mathrm{sh}\,\alpha_0}{\mathrm{ch}\,\alpha_0-\cos\beta_0},$$

$$z=\frac{c\sin\beta}{\mathrm{ch}\,\alpha-\cos\beta},\quad z_0=\frac{c\sin\beta_0}{\mathrm{ch}\,\alpha_0-\cos\beta_0},$$

$$\left(\varepsilon_0=\frac{1}{2},\quad \varepsilon_m=1,\quad m=1,2,\ldots\right).$$

u_0 is the potential of the point sources, and $Q_p(x)$ is a Legendre function of second kind.

The method of separation of variables leads to the following set of particular solutions:

$$u=(2\,\mathrm{ch}\,\alpha-2\cos\beta)^{\frac{1}{2}}\left[M_m(\nu)P^{-m}_{-\frac{1}{2}+i\nu}(\mathrm{ch}\,\alpha)+N_m(\nu)P^{-m}_{-\frac{1}{2}+i\nu}(-\mathrm{ch}\,\alpha)\right]\cdot\left[A_m(\nu)\,\mathrm{ch}\,\nu\beta+B_m(\nu)\,\mathrm{sh}\,\nu\beta\right]\cos m\varphi. \tag{27}$$

We shall seek the potential function of the secondary field in the form [1]:

$$u_1=(2\,\mathrm{ch}\,\alpha-2\cos\beta)^{\frac{1}{2}}\cdot\sum_{m=0}^{\infty}\varepsilon_m\cos m\varphi\int_0^{\infty}\left[A_m(\nu)\,\mathrm{ch}\,\nu\beta+B_m(\nu)\,\mathrm{sh}\,\nu\beta\right]P^{-m}_{-\frac{1}{2}+i\nu}(\mathrm{ch}\,\alpha)\,d\nu. \tag{28}$$

$$(\varepsilon_0 = \frac{1}{2}, \quad \varepsilon_m = 1, \quad m = 1, 2, \ldots).$$

Substituting (26) and (28) into (25), we obtain

$$-\frac{\mathcal{J}\, Q_{m-\frac{1}{2}}(\ldots)_{\beta=\beta_2}}{4\pi^2\sigma c\sqrt{\frac{2\,\mathrm{sh}\,\alpha\,\mathrm{sh}\,\alpha_0}{\mathrm{ch}\,\alpha_0 - \cos\beta_0}}} = \int_0^\infty [A_m(\nu)\,\mathrm{ch}\,\nu\beta_2 + B_m(\nu)\,\mathrm{sh}\,\nu\beta_2]\, P^{-m}_{-\frac{1}{2}+i\nu}(\mathrm{ch}\,\alpha)\, d\nu,$$

$$-\frac{\mathcal{J}\, Q_{m-\frac{1}{2}}(\ldots)_{\beta=\beta_1+2\pi}}{4\pi^2\sigma c\sqrt{\frac{2\,\mathrm{sh}\,\alpha\,\mathrm{sh}\,\alpha_0}{\mathrm{ch}\,\alpha_0 - \cos\beta_0}}} = \int_0^\infty [A_m(\nu)\,\mathrm{ch}\,\nu(\beta_1+2\pi) + B_m(\nu)\,\mathrm{sh}\,\nu(\beta_1+2\pi)]\, P^{-m}_{-\frac{1}{2}+i\nu}(\mathrm{ch}\,\alpha)\, d\nu. \tag{29}$$

Multiplying both sides of equations (29) by $\nu\,\mathrm{th}\,\nu\pi\,\frac{\Gamma(\frac{1}{2}+m+i\nu)\,\Gamma(\frac{1}{2}+m+i\nu)}{\Gamma(\frac{1}{2}+i\nu)\,\Gamma(\frac{1}{2}-i\nu)}\cdot P^{-m}_{-\frac{1}{2}+i\nu}(\mathrm{ch}\,\alpha)\,\mathrm{sh}\,\alpha$ and integrating with respect to α between the limits 0 and ∞, we obtain, on the basis of (13), the following algebraic system:

$$A_m(\nu)\,\mathrm{ch}\,\nu\beta_2 + B_m(\nu)\,\mathrm{sh}\,\nu\beta_2 = I^{(1)}_m,$$

$$A_m(\nu)\,\mathrm{ch}\,\nu(\beta_1+2\pi) + B_m(\nu)\,\mathrm{sh}\,\nu(\beta_1+2\pi) = I^{(2)}_m, \tag{30}$$

where

$$I^{(1)}_m = -\frac{\nu\,\mathrm{th}\,\nu\pi\,\frac{\Gamma(\frac{1}{2}+m+i\nu)\,\Gamma(\frac{1}{2}+m-i\nu)}{\Gamma(\frac{1}{2}+i\nu)\,\Gamma(\frac{1}{2}-i\nu)}\,\mathcal{J}}{4\pi^2\sigma c\sqrt{\frac{2\,\mathrm{sh}\,\alpha_0}{\mathrm{ch}\,\alpha_0-\cos\beta_0}}}\cdot\int_0^\infty Q_{m-\frac{1}{2}}(\ldots)_{\beta=\beta_2}\, P^{-m}_{-\frac{1}{2}+i\nu}(\mathrm{ch}\,\alpha)\sqrt{\mathrm{sh}\,\alpha}\, d\alpha,$$

$$I^{(2)}_m = -\frac{\nu\,\mathrm{th}\,\nu\pi\,\frac{\Gamma(\frac{1}{2}+m+i\nu)\,\Gamma(\frac{1}{2}+m-i\nu)}{\Gamma(\frac{1}{2}+i\nu)\,\Gamma(\frac{1}{2}-i\nu)}\,\mathcal{J}}{4\pi^2\sigma c\sqrt{\frac{2\,\mathrm{sh}\,\alpha_0}{\mathrm{ch}\,\alpha_0-\cos\beta_0}}}\cdot\int_0^\infty Q_{m-\frac{1}{2}}(\ldots)_{\beta=\beta_1+2\pi}\, P^{-m}_{-\frac{1}{2}+i\nu}(\mathrm{ch}\,\alpha)\sqrt{\mathrm{sh}\,\alpha}\, d\alpha. \tag{31}$$

From the system (30) we determine the coefficients

$$A_m(\nu) = \frac{\mathrm{sh}\,\nu(2\pi+\beta_1)\, I^{(1)}_m - \mathrm{sh}\,\nu\beta_2\, I^{(2)}_m}{\mathrm{sh}\,\nu(2\pi+\beta_1-\beta_2)},$$

$$B_m(\nu) = \frac{-\mathrm{ch}\,\nu(\beta_1+2\pi)\, I^{(1)}_m + \mathrm{ch}\,\nu\beta_2\, I^{(2)}_m}{\mathrm{sh}\,\nu(2\pi+\beta_1-\beta_2)}. \tag{32}$$

The desired potential function has the form

$$U = U_0 + U_1 = \frac{\mathcal{J}}{4\pi\sigma\sqrt{r^2+r_0^2-2rr_0\cos\varphi+(z-z_0)^2}} + (2\,\mathrm{ch}\,\alpha - 2\cos\beta)^{\frac{1}{2}}\sum_{m=0}^{\infty}\varepsilon_m\cos m\varphi\,\cdot$$

$$\cdot \int_0^\infty \frac{sh\,\nu(2\pi+\beta_1-\beta)\, I_m^{(1)} + sh\,\nu(\beta-\beta_2)\, I_m^{(2)}}{sh(2\pi+\beta_1-\beta_2)\nu} P^{-m}_{-\frac{1}{2}+i\nu}(ch\,\alpha)\,d\nu, \; (\varepsilon_0 = \tfrac{1}{2}, \quad \varepsilon_m = 1; \quad m = 1, 2, \ldots), \tag{33}$$

where $I_m^{(1)}$ and $I_m^{(2)}$ are defined by formulas (31).

Using the asymptotic representations for associated Legendre functions, it is possible to show that the integrals in (18) and (33) have the proper convergence properties and that formulas (18) and (33) actually give solutions of the problem in question.

Conclusions

The boundary value problems for the conducting hyperboloid and the ideally conducting lens in electrical prospecting lead to solution of the Laplace equation in degenerate ellipsoidal and toroidal coordinate systems.

Solutions are obtained in the form of Fourier series of integrals containing associated Legendre functions of first kind and their derivatives.

For the actual computation of the potential functions of these problems, it is necessary to tabulate the real functions $P^{-m}_{-\frac{1}{2}+i\nu}(x)$ and $P^{-m'}_{-\frac{1}{2}+i\nu}(x)$ for real values of x and the parameter $\nu > 0$.

The author wishes to thank V. M. Babich for reading this work and making a number of valuable suggestions.

LITERATURE CITED

1. Lebedev, N. N., Special Functions and Their Applications [in Russian] (1953).
2. Nikolaev, B. G., "The expansion of an arbitrary function in terms of an integral of associated Legendre functions of first kind with complex index," this volume, p. 45.
3. Glyuzman, A. M., "Solution of the boundary value problem for the hyperboloid of revolution in electrical prospecting," Izv. Akad. Nauk SSSR, No. 5 (1961).

APPLICATION OF THE LAPLACE METHOD TO THE CONSTRUCTION OF SOLUTIONS OF THE HELMHOLTZ EQUATION

B. G. Nikolaev

A technique closely related to the theory of integral transforms is used in many problems of mathematical physics. The method of integral transforms makes it possible to successfully study processes of wave propagation and diffraction [1-9]; it is widely used in solving problems of the theory of heat conduction [10-14] and also in a number of other cases [15-17]. However, together with the well-known advantages of this method, there are also shortcomings which appear in the solution of boundary value problems of mathematical physics with complicated conditions of the boundary. This latter fact is the reason that integral transforms have been used mainly in problems with time variables, while problems with spatial variables have been little considered by this method. Because of this, the necessity has long been felt of working out methods of integral transforms which take into account not only the particular features of the processes described but also the shape of the region in which these processes take place. Such methods have been proposed in [18-20]. It must be noted that to obtain new transforms, as a rule, is not an elementary problem. Moreover, in choosing a transform for some group of problems there is always a certain amount of freedom; this can either restrict the possibilities for applying the transform chosen or, on the other hand, extend them.

In the present paper we shall not be concerned with the general aspect of the questions mentioned, but we shall rather show by specific examples how a transform should be chosen, starting from the particular conditions of the problem under consideration.

The content of the present word was suggested by the papers of G. D. Malyuzhenets on the problem of wave propagation and diffraction in angular plane regions. The first such problems in this direction were successfully treated by Sommerfeld [21]. Many people after Sommerfeld improved the mathematical side of the method. However, until recently the method of Sommerfeld had not been essentially improved, since there was no known inversion formula for the integral transform used in this method. This gap was filled only after the work of G. D. Malyuzhenets [22-25], and Sommerfeld's method then assumed its proper place among other methods [29-32] used in solving diffraction problems in angular regions. One must, however, not think that Sommerfeld's method leads directly to the result in all cases of diffraction from corners. There is an entire series of problems which are naturally considered in another manner: for example, problems with multilayer angular regions and elastic media. One such method is proposed in the present paper. The method proposed makes it possible to solve in the same manner all previously solved problems of wave diffraction by angular plane regions and also affords the possibility of considering other more complicated problems. To illustrate the possibilities of this method, we consider the Neumann problem for stationary wave diffraction from an angular plane region in the case of a point source located on one of the edges. In subsequent publications we intend to consider the analogous problem for stationary wave diffraction from a half space separated by an inclined boundary into two angular regions with different wave propagation speeds.

§1. A Class of Solutions of the Helmholtz Equation Used in Problems of Wave Diffraction from Angular Regions

The object of the present section is the construction and study of solutions of the equation

$$L[W]=r^2\frac{\partial^2 W}{\partial r^2}+r\frac{\partial W}{\partial r}+\kappa^2 r^2 W+\frac{\partial^2 W}{\partial \varphi^2}=0 \tag{1.1}$$

for real independent variables r and φ and a complex parameter κ.

Since we intend to eventually apply the results to problems of wave diffraction by angular plane regions, we shall seek solutions of (1.1) in the form

$$W=\frac{1}{2\pi i}\int_C v(p,\varphi;\kappa)e^{pr}dp\ , \tag{1.2}$$

where $v(p,\varphi;\kappa)$ is an unknown function of the complex variable $p=\sigma+i\tau$ and the real variable φ, and c is a contour of integration which we shall subsequently specify.

Substituting (1.2) into (1.1) and integrating by parts, we obtain

$$2\pi i\, L[W]=\int_C M[v]e^{pr}dp-\Delta\Big|_C\ , \tag{1.3}$$

where

$$M[v]=\frac{\partial^2}{\partial p^2}(p^2+\kappa^2)v-\frac{\partial}{\partial p}(pv)+\frac{\partial^2 v}{\partial \varphi^2}\ ,$$

$$\Delta\Big|_C=\left\{\left[(p^2+\kappa^2)\left(\frac{\partial v}{\partial p}-rv\right)+pv\right]e^{pr}\right\}_C\ . \tag{1.4}$$

We remark that in obtaining (1.3) and (1.4) we started from the assumption that it was possible to interchange the order of the operation L and the operation of integration over the contour C. The justification for this must be checked after formally constructing a solution.

We now require that the following condition be satisfied on the contour C:

$$\Delta\Big|_C=0\ . \tag{1.5}$$

Equation (1.1) will now be satisfied if

$$M[v]=0\ . \tag{1.6}$$

Putting

$$p=\kappa\zeta,\qquad v(p,\varphi;\kappa)=\frac{U(\zeta,\varphi;\kappa)}{\kappa\sqrt{1+\zeta^2}}\ , \tag{1.7}$$

where U is a solution of the equation

$$(\zeta^2+1)\frac{\partial^2 U}{\partial \zeta^2}+\zeta\frac{\partial U}{\partial \zeta}+\frac{\partial^2 U}{\partial \varphi^2}=0, \tag{1.8}$$

and going over to the notation

$$\zeta=\operatorname{sh}\psi,\qquad U(\zeta,\varphi;\kappa)=\eta(\psi,\varphi;\kappa)\ , \tag{1.9}$$

we obtain

$$\frac{\partial^2\eta}{\partial\varphi^2}+\frac{\partial^2\eta}{\partial\psi^2}=0. \tag{1.10}$$

The general solution of (1.10) has the form

$$\eta = \eta_1(\varphi + \varphi_1 - i\psi) - \eta_2(\varphi - \varphi_2 + i\psi), \tag{1.11}$$

where the $\eta_j(z)$ are arbitrary functions of the complex variable z, and φ_1 and φ_2 are certain constants which will be specified when we consider the radiation condition.

Returning to the previous notation, we have

$$v(p, \varphi; \kappa) = \frac{\eta_1(\varphi + \varphi_1 - i\,\mathrm{arsh}\frac{p}{\kappa}) - \eta_2(\varphi - \varphi_2 + i\,\mathrm{arsh}\frac{p}{\kappa})}{\sqrt{p^2 + \kappa^2}}. \tag{1.12}$$

§2. The Choice of the Contour of Integration C in a Transform with a Laplace Kernel

Formula (1.12) specifies the general form of functions $v(p, \varphi; \kappa)$, in the transform (1.2). It remains to clarify the question of the choice of the contour C in the integral (1.2).

To be specific, we shall assume that the parameter κ in (1.1) is real and positive. The expressions $\mathrm{arsh}\frac{p}{\kappa}$ and $\sqrt{p^2+\kappa^2}$, in (1.12) have branch points $p = \pm i\kappa$ and $p = \infty$ in the (p) plane. In order to make $\mathrm{arsh}\frac{p}{\kappa}$ and $\sqrt{p^2+\kappa^2}$ single-valued functions, we introduce cuts in the (p) plane as shown in Fig. 1.

In the cut (p) plane we choose those branches of $\mathrm{arsh}\frac{p}{\kappa}$ and $\sqrt{p^2+\kappa^2}$, for which

$$-\frac{\pi}{2} \leqslant \mathrm{Im}\,\mathrm{arsh}\frac{p}{\kappa} \leqslant \frac{\pi}{2}, \quad \left|\mathrm{Re}\,\mathrm{arsh}\frac{p}{\kappa}\right| < \infty, \tag{2.1}$$

$$\arg\sqrt{p^2+\kappa^2} = 0 \quad \text{for} \quad \mathrm{Im}\,p = 0.$$

To determine the values of the function η from (1.11) at points of the complex (p) plane it is necessary to know the behavior of the function

$$z = \varphi_1 - i\,\mathrm{arsh}\frac{p}{\kappa} = x + iy \tag{2.2}$$

on this plane. From the simplest properties of conformal mapping and conditions (2.1) it is not difficult to determine the correspondence of regions for the variables p and z related by (2.2). Such correspondence is illustrated in Figs. 2 and 3, while the relation of the characteristic points of the function (2.2) is shown in Table 1.

Figures 2 and 3 give a graphic illustration of the correspondence between the regions when the cuts in the p plane are deformed.

Recalling formulas (1.4) and (1.12), we write condition (1.5) in the form

$$\Delta\Big|_C = \sqrt{p^2+\kappa^2}\left(\frac{\partial\eta}{\partial p} - r\eta\right)e^{pr}\Big|_C = 0, \tag{2.3}$$

where

$$\eta = \eta_1(\varphi + \varphi_1 - i\,\mathrm{arsh}\frac{p}{\kappa}) - \eta_2(\varphi - \varphi_2 + i\,\mathrm{arsh}\frac{p}{\kappa}). \tag{2.4}$$

TABLE 1

p	$-i\kappa$	0	$i\kappa$
z	$\varphi_1 - \frac{\pi}{2}$	φ_1	$\varphi_1 + \frac{\pi}{2}$

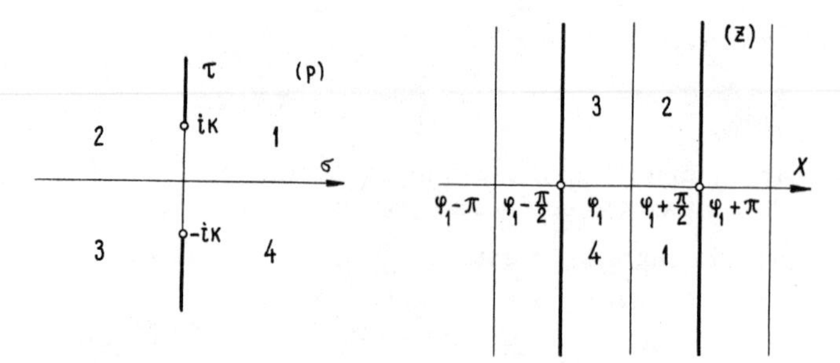

Fig. 1

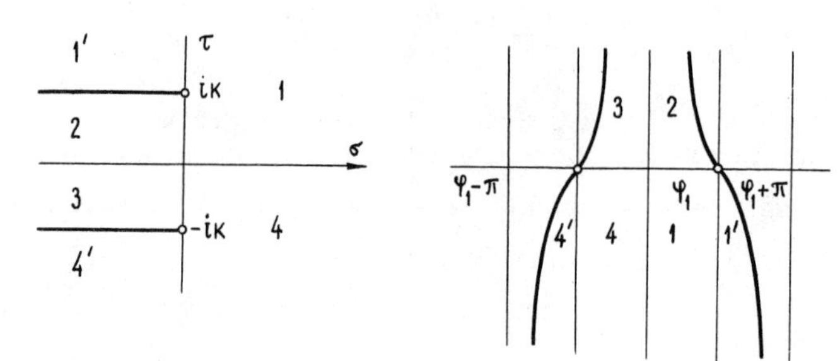

Fig. 2

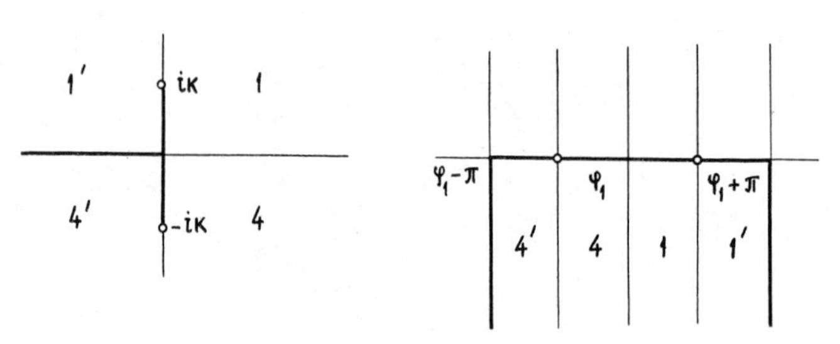

Fig. 3

Denoting by C_1 and C_2 the two contours shown in Fig. 4, it is easily seen that for a broad class of functions* the integral of (1.2), (1.12) converges on each of these paths for $\tau > 0$ and represents two linearly independent solutions of equation (1.1).

For regions $\varphi = const$ it is convenient to take the usual Mellin contour $\mathrm{Re}\, p = \sigma_0 > 0$ as the path of integration in (1.2). Condition (2.3) will hereby be fulfilled if it is required that

$$\sqrt{p^2+\kappa^2}\, e^{2p\tau} \frac{\partial}{\partial p}\left(e^{-p\tau}\eta\right)\Big|_{p=\sigma_0-i\infty}^{p=\sigma_0+i\infty} = 0,$$

$$\eta = \eta_1\left(\varphi+\varphi_1 - i\,\mathrm{arsh}\frac{p}{\kappa}\right) - \eta_2\left(\varphi-\varphi_2+i\,\mathrm{arsh}\frac{p}{\kappa}\right),$$

or, what is equivalent,

$$e^{2i\kappa\tau\sin(z-\varphi_1)}\frac{\partial}{\partial z}\left(e^{-i\kappa\tau\sin(z-\varphi_1)}\eta\right)\Big|_{z=\varphi_1-\frac{\pi}{2}+\varepsilon-i\infty}^{z=\varphi_1+\frac{\pi}{2}-\varepsilon-i\infty} = 0,$$

$$\eta = \eta_1(\varphi+z)-\eta_2(\varphi+\varphi_1-\varphi_2-z), \quad \varepsilon>0.$$

*For example, functions having polynomial growth at infinity belong to this class.

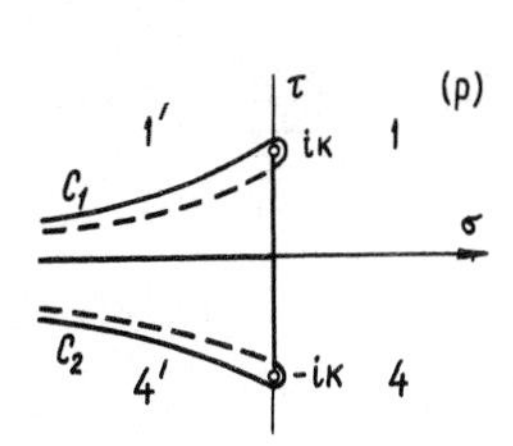

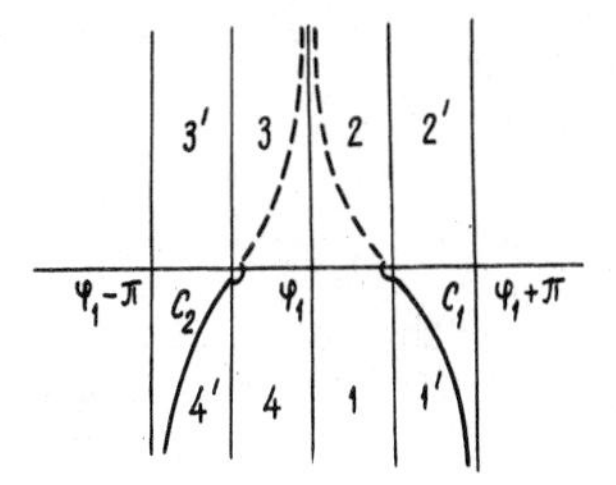

Fig. 4

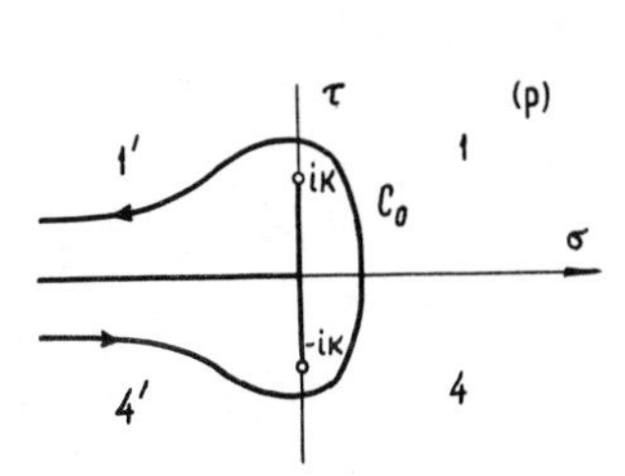

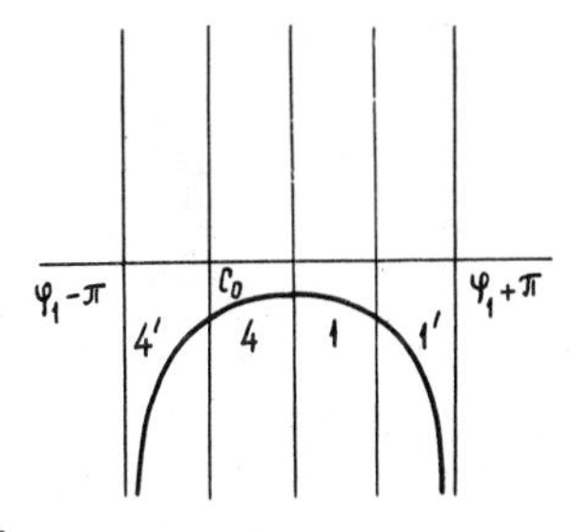

Fig. 5

If conditions (2.6) are not satisfied, then the contour shown in Fig. 5 should be taken as the path of integration C in (1.2).

§3. The Radiation Condition

We now consider questions related to the radiation principle. According to this principle, the total wave fields in diffraction problems with angular regions should not contain oscillations which have the character of cylindrical waves coming in from infinity.

We shall assume that the time dependence of the wave fields in the problems under consideration has the form $w(r,\varphi)\exp(-i\omega t)$, where $W(r,\varphi)$ is defined by the integral (1.2), (1.12) taken over the contour $\mathrm{Re}\, p=\sigma_0>0$.

By deforming the contour $\mathrm{Re}\, p=\sigma_0>0$ into the left half plane (the possibility of such a deformation must be checked after the formal construction of the solution), the evaluation of integral (1.2) can be reduced to evaluation of integrals of the same form but taken over the contours C' and C'' which go around the branch points (see Fig. 6).

Computations show that the integrals taken over the contours C' and C'' correspond respectively to outgoing and incoming cylindrical waves. If in (1.12) we put

$$\eta_1(z)=\eta_2(z)=\eta(z), \qquad \varphi_1=\varphi_2=\frac{\pi}{2}, \tag{3.1}$$

then it turns out that the function $v(p,\varphi;\kappa)$ has a removable branch at the point $p=-i\kappa$, and the integral (1.2), (1.12), (3.1) over the contour C'' is zero.

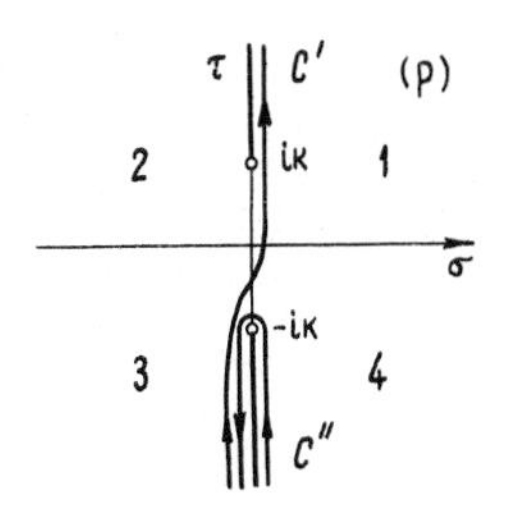

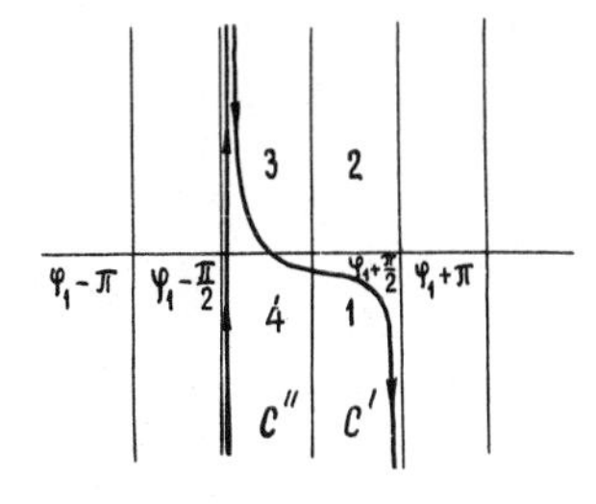

Fig. 6

Indeed, let us denote by

$$z_- = \frac{\pi}{2} - i\,\mathrm{arsh}\,\frac{p}{\kappa} = \frac{\pi}{2} - i\ln\left(\frac{p}{\kappa} + \sqrt{1 + \frac{p^2}{\kappa^2}}\right)$$

the value of the function z on the left side of the branch cut $(-i\kappa, -i\infty)$. Then on passing to the right side of this cut the function will assume the form:

$$z_+ = \frac{\pi}{2} - i\ln\left(\frac{p}{\kappa} - \sqrt{1 + \frac{p^2}{\kappa^2}}\right) = \frac{\pi}{2} - i\ln\frac{e^{-i\pi}}{\left(\frac{p}{\kappa} + \sqrt{1 + \frac{p^2}{\kappa^2}}\right)} = -\frac{\pi}{2} + i\ln\left(\frac{p}{\kappa} + \sqrt{1 + \frac{p^2}{\kappa^2}}\right) = -z_-$$

From this relation and from formula (1.12) in which $\eta_1 = \eta_2 = \eta$ and $\varphi_1 = \varphi_2 = \frac{\pi}{2}$, it follows that the function $v(p, \varphi; \kappa)$ does not change its value on passing through the cut $(-i\kappa, -i\infty)$. It can therefore be asserted that under condition (3.1) the integral (1.2), (1.12) over the contour C'' is zero.

As for the analogous integral along the contour c', under the conditions above it is not zero, and with the time factor $\exp(-i\omega t)$ it satisfies the radiation condition.

On the basis of the considerations above, the integral

$$W(r, \varphi, t) = \frac{e^{-i\omega t}}{2\pi i} \int_{\sigma_0 - i\infty}^{\sigma_0 + i\infty} \frac{\eta\left(\varphi + \frac{\pi}{2} - i\,\mathrm{arsh}\frac{p}{\kappa}\right) - \eta\left(\varphi - \frac{\pi}{2} + i\,\mathrm{arsh}\frac{p}{\kappa}\right)}{\sqrt{p^2 + \kappa^2}}\, e^{pr}\, dp \tag{3.2}$$

under condition (2.5) represents the general form of wave fields satisfying the radiation condition in problems of diffraction of waves from plane angular regions.

§4. An Expression for the Normal Derivative of the Desired Wave Function

To solve the Neumann problem for the case of a corner and also for problems with multilayer angular regions, it is necessary to operate with an expression for the normal derivative of the desired wave field. Such an expression is easily obtained from (3.2) by integrating by parts. Here

$$\frac{1}{r}\frac{\partial W}{\partial \varphi} = \frac{e^{-i\omega t}}{2\pi i} \int_{\sigma_0 - i\infty}^{\sigma_0 + i\infty} \frac{\eta'\left(\varphi + \frac{\pi}{2} - i\,\mathrm{arsh}\frac{p}{\kappa}\right) - \eta'\left(\varphi - \frac{\pi}{2} + i\,\mathrm{arsh}\frac{p}{\kappa}\right)}{\sqrt{p^2 + \kappa^2}} \frac{e^{pr}}{r}\, dp = \frac{e^{-i\omega t}}{2\pi i^2} \int_{\sigma_0 - i\infty}^{\sigma_0 + i\infty} \left[\eta\left(\varphi + \frac{\pi}{2} - i\,\mathrm{arsh}\frac{p}{\kappa}\right) + \eta\left(\varphi - \frac{\pi}{2} + i\,\mathrm{arsh}\frac{p}{\kappa}\right)\right] e^{pr}\, dp, \tag{4.1}$$

if

$$\left[\eta\left(\varphi + \frac{\pi}{2} - i\,\mathrm{arsh}\frac{p}{\kappa}\right) + \eta\left(\varphi - \frac{\pi}{2} + i\,\mathrm{arsh}\frac{p}{\kappa}\right)\right] e^{pr} \Big|_{\sigma_0 - i\infty}^{\sigma_0 + i\infty} = 0. \tag{4.2}$$

In the majority of practical cases the function $\eta(z)$ is such that condition (4.2) is satisfied, and the normal derivative of W can be represented in the form (4.1).

§5. Solution by the Laplace Method of the Neumann Problem for Stationary Diffraction Waves from an Angular Plane Region

The method developed in Sections 1-3 can be applied to solving a number of problems concerning diffraction of waves by a corner which are created by point sources located on the boundary of the angular region. One such problem is the Neumann problem for the Helmholtz equation.

The mathematical formulation of this problem is as follows. In the coordinate system (r, φ) there is given an angular region $(r > 0,\ 0 \leqslant \varphi \leqslant \alpha)$ in which wave processes are described by the equation

$$r^2 \frac{\partial^2 W}{\partial r^2} + r \frac{\partial W}{\partial r} + \kappa^2 r^2 W + \frac{\partial^2 W}{\partial \varphi^2} = 0 \tag{5.1}$$

with the condition that on the edges $\varphi = 0$ and $\varphi = \alpha$ the equations

$$\left(\frac{1}{r}\frac{\partial W}{\partial \varphi}\right)_{\varphi=0} = -\delta(r - r_0), \quad \left(\frac{1}{r}\frac{\partial W}{\partial \varphi}\right)_{\varphi=\alpha} = 0 \tag{5.2}$$

are satisfied, where

$$\delta(r) = \frac{1}{2\pi i}\int_{\sigma_0 - i\infty}^{\sigma_0 + i\infty} e^{pr}\, dp \tag{5.3}$$

is the Dirac symbol.

Under these conditions we seek a wave field $W(r,\varphi)\cdot exp(-i\omega t)$ which satisfies the radiation (see Section 3) and the requirement of being finite* at $r = 0$.

A solution of the problem formulated can be easily obtained on the basis of the principle of duality [33] from a solution for the analogous problem [34-35] for a point source located inside the angular region. Nevertheless, it is useful to consider such a problem in order to illustrate the possibilities of the method proposed.

According to Section 3, we seek a solution of the diffraction problem (5.1), (5.3) in the form (3.2) where the integration contour $Re\, p = \sigma_0 > 0$ is in the (p) plane cut along the segment $(i\kappa, i\infty)$. Substituting (3.2) into (5.1), (5.3), we obtain the system

$$\begin{gathered} \eta\left(\frac{\pi}{2} - i\, arsh\frac{p}{\kappa}\right) + \eta\left(-\frac{\pi}{2} + i\, arsh\frac{p}{\kappa}\right) = ie^{-pr}, \\ \eta\left(\alpha + \frac{\pi}{2} - i\, arsh\frac{p}{\kappa}\right) + \eta\left(\alpha - \frac{\pi}{2} + i\, arsh\frac{p}{\kappa}\right) = 0 \end{gathered} \tag{5.4}$$

for the function $\eta(z)$. Going over in (5.4) to the new variable $z = \frac{\pi}{2} - i\, arsh\frac{p}{\kappa}$ we have

$$\begin{gathered} \eta(z) + \eta(-z) = ie^{i\kappa r_0 \cos z}, \\ \eta(\alpha + z) + \eta(\alpha - z) = 0\,. \end{gathered} \tag{5.5}$$

We recall that the transformation $z = \frac{\pi}{2} - i\, arsh\frac{p}{\kappa}$ with the uniformization (2.1) takes a basic sheet of the (p) plane into the strip $0 < Re\, z < \pi$. Since the solution of the problem in question is defined only in a strip of the (z) plane, it is natural to put [36]

$$\eta(z) = \int_{-\infty}^{\infty} \chi(\xi) e^{-z\xi}\, d\xi\,. \tag{5.6}$$

Substituting (5.6) into (5.5) and using the identity [37]

$$ie^{i\kappa r_0 \cos z} = -\frac{1}{2}\int_{-\infty}^{\infty} e^{\frac{\pi\xi}{2}} H^{(1)}_{-i\xi}(\kappa r_0) e^{-z\xi}\, d\xi\,, \tag{5.7}$$

we obtain

$$\begin{gathered} \chi(\xi) + \chi(-\xi) = -\frac{1}{2} e^{\frac{\pi\xi}{2}} H^{(1)}_{-i\xi}(\kappa r_0), \\ e^{-\alpha\xi}\chi(\xi) + e^{\alpha\xi}\chi(-\xi) = 0\,. \end{gathered} \tag{5.8}$$

*For $r \to 0$ the function W must be determined from the conditions $\Delta W \simeq 0$, $\left(\frac{\partial W}{\partial \varphi}\right)_{\varphi=0} = 0$, $\left(\frac{\partial W}{\partial \varphi}\right)_{\varphi=\alpha} = 0$, and hence $W \sim c r^{\frac{\pi}{\alpha}} \cos\frac{\pi\varphi}{\alpha}$, $\frac{\partial W}{\partial r} \sim \frac{\pi}{\alpha} c r^{\frac{\pi}{\alpha}-1} \cos\frac{\pi\varphi}{\alpha}$.

Solving (5.8), we find

$$\eta(z)=-\frac{1}{4}\int_{-\infty}^{\infty}\frac{e^{\frac{\pi\xi}{2}}H_{-i\xi}^{(1)}(\kappa r_0)}{\operatorname{sh}\alpha\xi}e^{-(z-\alpha)\xi}d\xi. \tag{5.9}$$

Replacing ξ by $-\xi$ in (5.9), we obtain, in addition to (5.9), the following expression:

$$\eta(z)=\frac{1}{4}\int_{-\infty}^{\infty}\frac{e^{\frac{\pi\xi}{2}}H_{-i\xi}^{(1)}(\kappa r_0)}{\operatorname{sh}\alpha\xi}e^{+(z-\alpha)\xi}d\xi. \tag{5.10}$$

Formulas (5.9) and (5.10) imply

$$\eta=\eta(\varphi+z)-\eta(\varphi-z)=-\frac{1}{2}\int_{-\infty}^{\infty}\frac{e^{\frac{\pi\xi}{2}}H_{-i\xi}^{(1)}(\kappa r_0)\operatorname{sh}(\varphi-\frac{\alpha}{2})\xi}{\operatorname{ch}\frac{\alpha\xi}{2}}e^{-z\xi}d\xi. \tag{5.11}$$

Substituting the value of η from (5.11) into (3.2), interchanging the order of integration,* and using the identity [37]

$$J_\nu(\kappa r)=\frac{1}{2\pi i}\int_{\sigma_0-i\infty}^{\sigma_0+i\infty}\frac{e^{-\nu\operatorname{arsh}\frac{p}{\kappa}+pr}}{\sqrt{p^2+\kappa^2}}dp, \tag{5.12}$$

$$(\operatorname{Re}\nu>-\tfrac{1}{2}),$$

we obtain

$$W=-\frac{1}{2}\int_{-\infty}^{\infty}\frac{H_{-i\xi}^{(1)}(\kappa r_0)J_{-i\xi}(\kappa r)}{\operatorname{ch}\frac{\alpha\xi}{2}}\operatorname{sh}(\varphi-\frac{\alpha}{2})\xi\, d\xi. \tag{5.13}$$

Formula (5.13) defines the solution for $0<r\leqslant r_0$. Using the identity

$$H_{-i\xi}^{(1)}(\kappa r_0)J_{-i\xi}(\kappa r)=H_{-i\xi}^{(1)}(\kappa r)J_{-i\xi}(\kappa r_0)+\frac{H_{-i\xi}^{(2)}(\kappa r)H_{-i\xi}^{(1)}(\kappa r_0)-H_{-i\xi}^{(2)}(\kappa r_0)H_{-i\xi}^{(1)}(\kappa r)}{2}, \tag{5.14}$$

this solution can be continued to the region $r\geqslant r_0$. Indeed,

$$W=-\frac{1}{2}\int_{-\infty}^{\infty}\frac{H_{i\xi}^{(1)}(\kappa r)J_{-i\xi}(\kappa r_0)}{\operatorname{ch}\frac{\alpha\xi}{2}}\operatorname{sh}(\varphi-\frac{\alpha}{2})\xi\, d\xi-\frac{1}{4}\int_{-\infty}^{\infty}\frac{H_{-i\xi}^{(2)}(\kappa r)H_{-i\xi}^{(1)}(\kappa r_0)-H_{-i\xi}^{(2)}(\kappa r_0)H_{-i\xi}^{(1)}(\kappa r)}{\operatorname{ch}\frac{\alpha\xi}{2}}\operatorname{sh}(\varphi-\frac{\alpha}{2})\xi\, d\xi, \tag{5.15}$$

where the second integral is zero, as the integral of an odd function over a symmetric interval of integration. Thus,

$$W=\begin{cases}-\frac{1}{2}\int_{-\infty}^{\infty}\frac{H_{-i\xi}^{(1)}(\kappa r_0)J_{i\xi}(\kappa r)}{\operatorname{ch}\frac{\alpha\xi}{2}}\operatorname{sh}(\varphi-\frac{\alpha}{2})\xi\, d\xi, & (r\leqslant r_0),\\ -\frac{1}{2}\int_{-\infty}^{\infty}\frac{H_{-i\xi}^{(1)}(\kappa r)J_{-i\xi}(\kappa r_0)}{\operatorname{ch}\frac{\alpha\xi}{2}}\operatorname{sh}(\varphi-\frac{\alpha}{2})\xi\, d\xi, & (r\geqslant r_0).\end{cases} \tag{5.16}$$

*This interchange of integrals is justified here, since for the chosen branch of the function $\operatorname{arsh}\frac{p}{\kappa}$ the inner integral converges uniformly with respect to p and the outer integral exists.

In the case $0 < \varphi < \alpha$ the residue theorem can be applied to the integrals (5.16). This gives

$$W = \begin{cases} \frac{2\pi}{\alpha i} \sum_{m=0}^{\infty} (-1)^m H^{(1)}_{(2m+1)\frac{\pi}{\alpha}}(\kappa r_0) J_{(2m+1)\frac{\pi}{\alpha}}(\kappa r) \sin (2m+1)\frac{\pi}{\alpha}(\varphi - \frac{\alpha}{2}), (r \leq r_0), \\ \frac{2\pi}{\alpha i} \sum_{m=0}^{\infty} (-1)^m H^{(1)}_{(2m+1)\frac{\pi}{\alpha}}(\kappa r) J_{(2m+1)\frac{\pi}{\alpha}}(\kappa r_0) \sin (2m+1)\frac{\pi}{\alpha}(\varphi - \frac{\alpha}{2}), (r \geq r_0). \end{cases} \quad (5.17)$$

It is not hard to see that formulas (5.17) satisfy all the conditions of the problem posed. Indeed, in the case $(2m+1)\frac{\pi}{\alpha} \gg \kappa r_0$ we have the estimate

$$J_{(2m+1)\frac{\pi}{\alpha}}(\kappa r_0) H^{(1)}_{(2m+1)\frac{\pi}{\alpha}}(\kappa r) \sim \frac{i e^{-(2m+1)\frac{\pi}{\alpha} \ln \frac{r}{r_0}}}{(2m+1)\frac{\pi^2}{\alpha}}, \quad (5.18)$$

$$(r_0 \leq r),$$

which follows from Debye's second asymptotic representation. It is evident from this estimate that the series (5.17) admits an arbitrary number of differentiations under the sum for $r \neq r_0$ and, in particular, they satisfy the Helmholtz equation. Moreover, from the form of the expansions

$$H^{(1)}_{(2m+1)\frac{\pi}{\alpha}}(\kappa r) \sim \sqrt{\frac{2}{\pi \kappa r}}\, e^{i[\kappa r - (2m+1)\frac{\pi^2}{2\alpha} - \frac{\pi}{4}]}, \quad (\kappa r \gg 1),$$

$$J_{(2m+1)\frac{\pi}{\alpha}}(\kappa r) \sim \frac{(\kappa r)^{(2m+1)\frac{\pi}{\alpha}}}{2^{(2m+1)\frac{\pi}{\alpha}} \Gamma[1+(2m+1)\frac{\pi}{\alpha}]}, \quad (\kappa r < 1). \quad (5.19)$$

for cylindrical functions $H^{(1)}_p(z)$ and $J_p(z)$ it follows that the function W satisfies the radiation condition at infinity and the condition of being finite at $r = 0$.

Finally, on differentiating with respect to the variable φ, it is not hard to see that the boundary conditions on the planes $\varphi = 0$ and $\varphi = \alpha$ are satisfied.

With this, the process of justifying the solution obtained may be considered finished.

The author wishes to thank G. I. Petrashen' under whose supervision this work was completed.

LITERATURE CITED

1. Petrashen', G. I., Uch. Zap. Leningrad. Gos. Univ., No. 208 (1956).
2. Molotkov, I. A., Dokl. Akad. Nauk SSSR, Vol., 140, No. 3 (1961).
3. Shemyakin, E. I., Priklad. Matem. i Mekh., Vol., 22, No. 3 (1958).
4. Nikolaev, B. G., Problems in the Dynamical Theory of the Propagation of Seismic Waves [in Russian], Vol. 3, Izd. Len. Gos. Univ., Leningrad (1959).
5. Grinberg, G. A., Selected Problems in the Mathematical Theory of Electric and Magnetic Phenomena [in Russian], Izd. Akad. Nauk SSSR (1948).
6. Kontorovich, M. I., and Lebedev, N. N., Zh. Tekh. Fiz., Vol. 8 (1938).
7. Kontorovich, M. I., and Lebedev, N. N., Zh. Tekh. Fiz., Vol. 9 (1939).
8. Malyuzhenets, G. D., Akust. Zh., No. 2 (1955).
9. Malyuzhenets, G. D., Akust. Zh., No. 3 (1955).
10. Doetch, G., Math. Ann., Vol. 112 (1935).
11. Tranter, C., Int. Trans. in Math. Phys. (1951).
12. Sneddon, J., Fourier Transforms, McGraw-Hill, New York (1951).
13. Lykov, A. V., Heat Conduction in Nonstationary Processes [in Russian] (1948).
14. Smirnov, M. S., Application of Integral Transforms to the Solution of Problems in the Theory of Molecular Transport [in Russian], Dissertation (1955).
15. Smythe, W. R., Static and Dynamic Electricity, McGraw-Hill, New York (1967).

16. Agranovich, Z. S., and Povsner, A. Ya., Application of Operational Methods ot the Solution of Certain Problems in Mathematical Physics [in Russian], Kharkov.
17. Titchmarsh, E., Eigenfunction Expansions Associated with Second-Order Differential Equations, Clarendon Press, Oxford (1950).
18. Ivanov, A. V., The Finite Laplace Transform and Operational Calculus in Several Variables [in Russian], Sb. tr. MT i pp (1956).
19. Meijer, C., Proc. Ned. Acad. Wetensch., Vol. 43 (1940).
20. Akhiezer. N. I., Lectures on the Theory of Approximation [in Russian], Moscow (1947).
21. Sommerfeld, A., Math. Ann., Vol. 47 (1896).
22. Epstein, Math. Ann., Vol. 47 (1896)
23. Macdonald, H. M., Proc. London Math. Soc., Series 2, Vol. 14, Part 5, pp. 410-427 (1914).
24. Reich, Ann. Phys., Vol. 37 (1912).
25. Malyuzhenets, G. D., Dokl. Akad. Nauk SSSR, Vol. 110, No. 1 (1958).
26. Malyuzhenets, G. D., Dokl. Akad. Nauk SSSR, Vol. 118, No. 6 (1958).
27. Malyuzhenets, G. D., Dokl. Akad. Nauk SSSR, Vol. 121, No. 3 (1958).
28. Malyuzhenets, G. D., Dokl. Akad. Nauk SSSR, Vol. 146, No. 5 (1962).
29. Maye, A. B., "Die Beugung elastischer Wellen an der Halbebene," Z. angew. Math. Mech. Vo. 33 (1953).
30. Sobolev, S. L., Tr. Steklov, Inst. Akad. Nauk SSSR, No. 41 (1934).
31. Rubinovich, A., Math. Ann., Vol. 96 (1927).
32. Tesser, R., Byull. Pol'skoi Akad. Nauk, Vol. 4, No. 7 (1956).
33. Babich, V. M., Problems in the Dynamical Theory of the Propagation of Seismic Waves [in Russian], Vol. 6, Izd. Len. Gos. Univ., Leningrad (1962).
34. Petrashen', G. I., Nikolaev, B. G., and Kouzov, D. P., Uch. Zap. Len. Gos. Univ., No. 246 (1958).
35. Nikolaev, B. G., and Vasil'eva, M. V., Uch. Zap. Len. Gos. Univ., No. 246 (1958).
36. van der Pol, B., Operational Calculus Based on the Two-Sided Laplace Integral, Cambridge Univ. Press, Cambridge (1950).
37. Gradshteyn, I. S., and Ryzhik, I. M., Table of Integrals, Series, and Products [Engl. transl.], Academic Press, New York (1965).

THE PROBLEM OF CONSTRUCTING SOLUTIONS OF THE NEUMANN PROBLEM FOR THE STATIONARY DIFFRACTION OF WAVES FROM A HALF SPACE SEPARATED BY AN INCLINED BOUNDARY INTO TWO ANGULAR REGIONS WITH DIFFERENT WAVE PROPAGATION SPEEDS

B. G. Nikolaev

This paper is devoted to the application of the method of integral transforms [1] to one of the most difficult areas of diffraction, which include problems of multilayer angular regions and elastic media. One of the typical problems of this class is the Neumann problem for a half space separated by an inclined rectilinear boundary into two angular regions with different wave propagation speeds. The mathematical formulation of this problem is as follows.

In the coordinate system (r, φ) are given continuous angular regions $q=1$ $(r \geq 0,\ 0 \leq \varphi \leq \alpha)$ and $q=2$ $(r \geq 0,\ \alpha < \varphi < 2\pi)$ in which wave processes are defined by the equations

$$r^2 \frac{\partial^2 W_q}{\partial r^2} + r \frac{\partial W_q}{\partial r} + K_q^2 r^2 W_q + \frac{\partial^2 W_q}{\partial \varphi^2} = 0 \tag{1}$$

under the condition

$$\begin{aligned} &\left(\frac{1}{r}\frac{\partial W_1}{\partial \varphi}\right)_{\varphi=0} = f(r), \qquad (W_1)_{\varphi=\alpha} = \rho (W_2)_{\varphi=\alpha}, \\ &\left(\frac{1}{r}\frac{\partial W_2}{\partial \varphi}\right)_{\varphi=\pi} = 0, \qquad \left(\frac{\partial W_1}{\partial \varphi}\right)_{\varphi=\alpha} = \left(\frac{\partial W_2}{\partial \varphi}\right)_{\varphi=\alpha}, \end{aligned} \tag{2}$$

where ρ is a constant characterizing the conditions of contact of media $q=1$ and $q=2$, and $f(r)$ is a function defined by the type of action.

Under these conditions wave fields $W_q\, exp(-i\omega t)$ are sought which satisfy the radiation condition at infinity and the condition of being finite at $r=0$.

According to [1], a solution of the problem just posed should be sought in the form

$$W_q = \frac{1}{2\pi i} \int_{\sigma_0 - i\infty}^{\sigma_0 + i\infty} \frac{\eta_q(\varphi + \frac{\pi}{2} - i\,arsh\frac{p}{K}) - \eta_q(\varphi - \frac{\pi}{2} + i\,arsh\frac{p}{K})}{\sqrt{p^2 + K_q^2}} e^{pr} dp, \tag{3}$$

$$(q = 1, 2),$$

where the contour $Re\, p = \sigma_0 > 0$ lies in the (p) plane to the right of the cut $(i\kappa,\ i\infty)$.

Putting

$$f(\tau) = \frac{1}{2\pi i}\int_{\sigma_0 - i\infty}^{\sigma_0 + i\infty} F(p)e^{p\tau}dp \tag{4}$$

and using (3), we obtain from the boundary conditions (2) the following system:

$$\begin{gathered}
\eta_1\left(\frac{\pi}{2} - i\,arsh\frac{p}{\kappa_1}\right) + \eta_1\left(-\frac{\pi}{2} + i\,arsh\frac{p}{\kappa_1}\right) = F(p), \\
\frac{\eta_1\left(\alpha + \frac{\pi}{2} - i\,arsh\frac{p}{\kappa_1}\right) - \eta_1\left(\alpha - \frac{\pi}{2} + i\,arsh\frac{p}{\kappa_1}\right)}{\sqrt{p^2 + \kappa_1^2}} = \rho\,\frac{\eta_2\left(\alpha + \frac{\pi}{2} - i\,arsh\frac{p}{\kappa_2}\right) - \eta_2\left(\alpha - \frac{\pi}{2} + i\,arsh\frac{p}{\kappa_2}\right)}{\sqrt{p^2 + \kappa_2^2}}, \\
\eta_1\left(\alpha + \frac{\pi}{2} - i\,arsh\frac{p}{\kappa_1}\right) + \eta_1\left(\alpha - \frac{\pi}{2} + i\,arsh\frac{p}{\kappa_1}\right) = \eta_2\left(\alpha + \frac{\pi}{2} - i\,arsh\frac{p}{\kappa_2}\right) + \eta_2\left(\alpha - \frac{\pi}{2} + i\,arsh\frac{p}{\kappa_2}\right), \\
\eta_2\left(\pi + \frac{\pi}{2} - i\,arsh\frac{p}{\kappa_2}\right) + \eta_2\left(\pi - \frac{\pi}{2} + i\,arsh\frac{p}{\kappa_2}\right) = 0.
\end{gathered} \tag{5}$$

for determining the unknown functions $\eta_1(z)$ and $\eta_2(z)$.

Going over to the new variables $z = \frac{\pi}{2} - i\,arsh\frac{p}{\kappa_1}$ in (5), we have

$$\begin{aligned}
&\eta_1(z) + \eta_1(-z) = F(-i\kappa_1 \cos z), \\
&\eta_1(\alpha + z) - \eta_1(\alpha - z) = \rho P(z)\left[\eta_2(\alpha + \Phi) - \eta_2(\alpha - \Phi)\right], \\
&\eta_1(\alpha + z) + \eta_1(\alpha - z) = \eta_2(\alpha + \Phi) + \eta_2(\alpha - \Phi), \\
&\eta_2(\pi + \Phi) + \eta_2(\pi - \Phi) = 0,
\end{aligned} \tag{6}$$

where (see Appendix 1)

$$\begin{gathered}
\Phi = \arccos\frac{\kappa_1}{\kappa_2}\cos z, \qquad \Phi\Big|_{\kappa_1 = \kappa_2} = z, \\
P(z) = \frac{d\Phi}{dz} = \frac{\kappa_1 \sin z}{\sqrt{\kappa_2^2 - \kappa_1^2\cos^2 z}}, \quad \Phi(z) = -\Phi(-z), \\
\Phi(z + n\pi) = n\pi + \Phi(z), \qquad P(z) = P(-z), \\
P(z + n\pi) = P(z).
\end{gathered} \tag{7}$$

A solution of (6) is naturally represented in the form

$$\begin{gathered}
\eta_1(z) = \overset{-}{X}_1(z - \alpha) + \overset{+}{Y}_1(z - \alpha), \\
\eta_2(z) = \overset{-}{X}_2(z - \alpha) + \overset{+}{Y}_2(z - \alpha), \\
\overset{-}{X}_\nu(z) = -\overset{-}{X}_\nu(-z), \quad \overset{+}{Y}_\nu(z) = \overset{+}{Y}_\nu(-z).
\end{gathered} \tag{8}$$

Here

$$
\begin{aligned}
&X_1^-(z) = \rho P(z) X_2^-(\Phi), \\
&Y_1^+(z) = Y_2^+(\Phi), \\
&X_1^-(z-\alpha) - X_1^-(z+\alpha) + Y_1^+(z-\alpha) + Y_1^+(z+\alpha) = F(-ik, \cos z), \\
&X_2^-(z+\pi-\alpha) - X_2^-(z-\pi+\alpha) + Y_2^+(z+\pi-\alpha) + Y_2^+(z-\pi+\alpha) = 0.
\end{aligned} \tag{9}
$$

Putting in (9)

$$
\begin{aligned}
&X_1^-(z) = v.p.\int_{-\infty}^{\infty} \frac{\chi_1^+(\zeta)}{sh\,\alpha\zeta} e^{-z\zeta} d\zeta, \\
&Y_1^+(z) = \int_{-\infty}^{\infty} \frac{\varkappa_1^+(\zeta)}{ch\,\alpha\zeta} e^{-z\zeta} d\zeta, \\
&X_2^-(z) = v.p.\int_{-\infty}^{\infty} \frac{\Omega_2^+(\zeta)}{sh(\pi-\alpha)\zeta} e^{-z\zeta} d\zeta, \\
&F(-ik, \cos z) = \int_{-\infty}^{\infty} \Psi^+(\zeta) e^{-z\zeta} d\zeta,
\end{aligned} \tag{10}
$$

we obtain

$$
\begin{aligned}
&\chi_1^+(\zeta) + \varkappa_1^+(\zeta) = \Psi^+(\zeta), \\
&v.p.\int_{-\infty}^{\infty} \frac{\chi_1^+(\zeta)}{sh\,\alpha\zeta} e^{-z\zeta} d\zeta = \rho P(z)\, v.p.\int_{-\infty}^{\infty} \frac{\Omega_2^+(\zeta)}{sh(\pi-\alpha)\zeta} e^{-\zeta\Phi(z)} d\zeta, \\
&\int_{-\infty}^{\infty} \frac{\varkappa_1^+(\zeta)}{ch\,\alpha\zeta} e^{-z\zeta} d\zeta = \int_{-\infty}^{\infty} \frac{\Omega_2^+(\zeta)}{ch(\pi-\alpha)\zeta} e^{-\zeta\Phi(z)} d\zeta,
\end{aligned} \tag{11}
$$

or

$$
\begin{aligned}
&\frac{\chi_1^+(\zeta)}{\rho\, sh\,\alpha\zeta} = v.p.\int_{-\infty}^{\infty} \frac{K_1(\zeta,\xi)\Omega_2^+(\xi)}{sh(\pi-\alpha)\xi} d\xi, \\
&\frac{\Psi^+(\zeta) - \chi_1^+(\zeta)}{ch\,\alpha\zeta} = \int_{-\infty}^{\infty} \frac{K_2(\zeta,\xi)\Omega_2^+(\xi)}{ch(\pi-\alpha)\xi} d\xi,
\end{aligned} \tag{12}
$$

where the generalized kernels are defined by the formulas

$$
\begin{aligned}
&P(z) e^{-\Phi(z)\zeta} = \int_{-\infty}^{\infty} K_1(\xi,\zeta) e^{-z\xi} d\xi, \\
&e^{-\Phi(z)\zeta} = \int_{-\infty}^{\infty} K_2(\xi,\zeta) e^{-z\xi} d\xi,
\end{aligned} \tag{13}
$$

$$K_1(\xi,\varkappa)=\frac{1}{2\pi i}\int\limits_{Re z=\sigma_0>0} P(z)e^{-\Phi(z)\varkappa+z\xi}\,dz,$$

$$K_2(\xi,\varkappa)=\frac{1}{2\pi i}\int\limits_{Re z=\sigma_0>0} e^{-\Phi(z)\varkappa+z\xi}\,dz.$$

Eliminating the function $\chi_1^+(\varkappa)$ in (12), we arrive at the equation

$$\Psi^+(\varkappa)=v.p.\int\limits_{-\infty}^{\infty}\Big[\frac{ch\,\alpha\varkappa}{ch(\pi-\alpha)\xi}K_2(\varkappa,\xi)+\frac{\rho\,sh\,\alpha\varkappa}{sh(\pi-\alpha)\xi}K_1(\varkappa,\xi)\Big]\Omega_2^+(\xi)\,d\xi$$

for determining the function $\Omega_2^+(\xi)$. This equation can be simplified if we consider the relation between the kernels (see Appendix 2)

$$K_1(\varkappa,\xi)=\frac{\varkappa}{\xi}K_2(\varkappa,\xi),\qquad K_2(\varkappa,\xi)=K_2(-\varkappa,-\xi), \tag{15}$$

and introduce a new function $Q_2^+(\varkappa)$ by

$$\Omega_2^+(\varkappa)=\frac{Q_2^+(\varkappa)}{\Delta^+(\varkappa,\varkappa)}, \tag{16}$$

where

$$\Delta^+(\varkappa,\xi)=\frac{ch\,\alpha\varkappa}{ch(\pi-\alpha)\xi}+\frac{\rho\varkappa\,sh\,\alpha\varkappa}{\xi\,sh(\pi-\alpha)\xi}. \tag{17}$$

Then

$$\Psi^+(\varkappa)=v.p.\int\limits_{-\infty}^{\infty}\frac{\Delta^+(\varkappa,\xi)}{\Delta^+(\xi,\xi)}K_2^+(\varkappa,\xi)Q_2^+(\xi)\,d\xi,\qquad K_2^+(\varkappa,\xi)=\frac{K_2(\varkappa,\xi)+K_2(-\varkappa,\xi)}{2}. \tag{18}$$

From the representation (see Appendix 2)

$$K_2(\varkappa,\xi)=\frac{1}{2\pi}\int\limits_{-\infty}^{\infty}\cos(z\varkappa-\xi\,\mathrm{arcch}\,\gamma\,ch\,z)\,dz \tag{19}$$

follow the limiting formulas

$$K_2^+(\varkappa,0)=\delta(\varkappa),\quad \text{for}\quad \gamma>0,\qquad K_2^+(\varkappa,\xi)=\delta(\varkappa-\xi),\quad \text{if}\quad \gamma=1, \tag{20}$$

where $\delta(z)$ denotes the Dirac function.

In the case $\gamma \geqslant 1$ (see Appendix 2) there exists a more general representation

$$K_2^+(\varkappa,\xi)=\Gamma^+(\xi,\gamma)\,\delta(\varkappa-\xi)+R^+(\varkappa,\xi;\gamma) \tag{21}$$

in which

$$\Gamma^+(\xi,\gamma)=\cos(\xi\ln\gamma),$$

$$R^+(\varkappa,\xi;\gamma)=-\frac{\xi\sin(\xi\ln\gamma)}{16}\left(1-\frac{1}{\gamma^2}\right)\frac{\varkappa-\xi}{sh\frac{\pi(\varkappa-\xi)}{2}}+\sum_{n=2}^{\infty}\frac{\cos(\xi\ln\gamma)a_n^+(\xi)+\sin(\xi\ln\gamma)b_n^-(\xi)}{4n!\,(2n-1)!}\,\frac{\varkappa-\xi}{sh\frac{\pi(\varkappa-\xi)}{2}}\,.$$

$$\cdot\prod_{k=1}^{n-1}\left[(\varkappa-\xi)^2+4k^2\right]-\frac{1}{2\pi}\int_{-\infty}^{\infty}\sin(\varkappa-\xi)z\cdot\sin\xi(z-\operatorname{arcch}\gamma\, ch z)\,dz\,. \tag{22}$$

Substituting (21) into (18) and using the notation

$$Q_2^+(\varkappa)=\frac{Q^+(\varkappa)}{\Gamma^+(\varkappa;\gamma)}\,, \tag{23}$$

we arrive at the equation

$$Q^+(\varkappa)+v.p.\int_{-\infty}^{\infty}\frac{\Delta^+(\varkappa,\xi)}{\Delta^+(\xi,\xi)}\,\frac{R^+(\varkappa,\xi;\gamma)}{\Gamma^+(\xi,\gamma)}\,Q^+(\xi)\,d\xi=\Psi^+(\varkappa) \tag{24}$$

for the function $Q^+(\varkappa)$. $\eta_1(z)$ and $\eta_2(z)$ are then obtained from the expressions

$$\eta_1(z)=v.p.\int_{-\infty}^{\infty}\frac{[sh(\pi-\alpha)\varkappa+\rho P(z-\alpha)\,ch(\pi-\alpha)\varkappa]\,Q^+(\varkappa)e^{-\varkappa\Phi(z-\alpha)}d\varkappa}{\Gamma^+(\varkappa;\gamma)[ch\alpha\varkappa\, sh(\pi-\alpha)\varkappa+\rho\, sh\alpha\varkappa\, ch(\pi-\alpha)\varkappa]},$$

$$\eta_2(z)=v.p.\int_{-\infty}^{\infty}\frac{[sh(\pi-\alpha)\varkappa+ch(\pi-\alpha)\varkappa]\,Q^+(\varkappa)e^{-\varkappa(z-\alpha)}d\varkappa}{\Gamma^+(\varkappa;\gamma)[ch\alpha\varkappa\, sh(\pi-\alpha)\varkappa+\rho\, sh\alpha\varkappa\, ch(\pi-\alpha)\varkappa]}\,. \tag{25}$$

This completes the formal construction of a solution of the system (6).

It remains to show that formulas (25) lead to the right boundary conditions on the planes $\varphi=\pi$ and $\varphi=\alpha$, and also that they define a certain class of functions $F(-ik_1\cos z)$ which correspond to the initial disturbance on the boundary $\varphi=0$.

From (25) we have

$$\eta_2(\pi+\Phi)+\eta_2(\pi-\Phi)=v.p.\int_{-\infty}^{\infty}\frac{[e^{-\varkappa\Phi}+e^{\varkappa\Phi}]\,Q^+(\varkappa)\,d\varkappa}{\Gamma^+(\varkappa;\gamma)[ch\alpha\varkappa\, sh(\pi-\alpha)\varkappa+\rho\, sh\alpha\varkappa\, ch(\pi-\alpha)\varkappa]}=0, \tag{26}$$

as the integral of an odd function over a symmetric interval of integration.

Further

$$\eta_2(\alpha+\Phi)+\eta_2(\alpha-\Phi) = v.p.\int_{-\infty}^{\infty} \frac{sh(\pi-\alpha)\xi\,[e^{-\xi\Phi}+e^{\xi\Phi}]\,\overset{+}{Q}(\xi)\,d\xi}{\overset{+}{\Gamma}(\xi;\gamma)[ch\alpha\xi\, sh(\pi-\alpha)\xi+\rho\, sh\alpha\xi\, ch(\pi-\alpha)\xi]}+$$

$$+v.p.\int_{-\infty}^{\infty} \frac{ch(\pi-\alpha)\xi\,[e^{-\xi\Phi}+e^{\xi\Phi}]\,\overset{+}{Q}(\xi)\,d\xi}{\overset{+}{\Gamma}(\xi;\gamma)[ch\alpha\xi\, sh(\pi-\alpha)\xi+\rho\, sh\alpha\xi\, ch(\pi-\alpha)\xi]} = I_1+I_2, \tag{27}$$

where $I_2 = 0$ for the same reason. On the other hand,

$$\eta_1(\alpha+z)+\eta_1(\alpha-z) = v.p.\int_{-\infty}^{\infty} \frac{sh(\pi-\alpha)\xi\,[e^{-\xi\Phi}+e^{\xi\Phi}]\,\overset{+}{Q}(\xi)\,d\xi}{\overset{+}{\Gamma}(\xi;\gamma)[ch\alpha\xi\, sh(\pi-\alpha)\xi+\rho\, sh\alpha\xi\, ch(\pi-\alpha)\xi]}+$$

$$+v.p.\int_{-\infty}^{\infty} \frac{\rho P(z)\, ch(\pi-\alpha)\xi\,[e^{-\xi\Phi}+e^{\xi\Phi}]\,\overset{+}{Q}(\xi)\,d\xi}{\overset{+}{\Gamma}(\xi;\gamma)[ch\alpha\xi\, sh(\pi-\alpha)\xi+\rho\, sh\alpha\xi\, ch(\pi-\alpha)\xi]} = I_1+I_3, \tag{28}$$

where $I_3 = 0$. Comparing formulas (27) and (28), it follows that

$$\eta_1(\alpha+z)+\eta_1(\alpha-z) = \eta_2(\alpha+\Phi)+\eta_2(\alpha-\Phi). \tag{29}$$

In precisely the same way

$$\eta_2(\alpha+\varphi)-\eta_2(\alpha-\varphi) = v.p.\int_{-\infty}^{\infty} \frac{ch(\pi-\alpha)\xi\,[e^{-\xi\Phi}-e^{\xi\Phi}]\,\overset{+}{Q}(\xi)\,d\xi}{\overset{+}{\Gamma}(\xi;\gamma)[ch\alpha\xi\, sh(\pi-\alpha)\xi+\rho\, sh\alpha\xi\, ch(\pi-\alpha)\xi]}+$$

$$+v.p.\int_{-\infty}^{\infty} \frac{sh(\pi-\alpha)\xi\,[e^{-\xi\Phi}-e^{\xi\Phi}]\,\overset{+}{Q}(\xi)\,d\xi}{\overset{+}{\Gamma}(\xi;\gamma)[ch\alpha\xi\, sh(\pi-\alpha)\xi+\rho\, sh\alpha\xi\, ch(\pi-\alpha)\xi]} = I_4+I_5, \tag{30}$$

where $I_5 = 0$ and

$$\eta_1(\alpha+z)-\eta_1(\alpha-z) = v.p.\int_{-\infty}^{\infty} \frac{\rho P(z)\, ch(\pi-\alpha)\xi\,[e^{-\xi\Phi}-e^{\xi\Phi}]\,\overset{+}{Q}(\xi)\,d\xi}{\overset{+}{\Gamma}(\xi;\gamma)[ch\alpha\xi\, sh(\pi-\alpha)\xi+\rho\, sh\alpha\xi\, ch(\pi-\alpha)\xi]}+$$

$$+v.p.\int_{-\infty}^{\infty} \frac{sh(\pi-\alpha)\xi\,[e^{-\xi\Phi}-e^{\xi\Phi}]\,\overset{+}{Q}(\xi)\,d\xi}{\overset{+}{\Gamma}(\xi;\gamma)[ch\alpha\xi\, sh(\pi-\alpha)\xi+\rho\, sh\alpha\xi\, ch(\pi-\alpha)\xi]} = \rho P(z)\, I_4+I_6, \tag{31}$$

where $I_6 = 0$. Therefore

$$\eta_1(\alpha+z)-\eta_1(\alpha-z) = \rho P(z)\,[\eta_2(\alpha+\Phi)-\eta_2(\alpha-\Phi)]. \tag{32}$$

Formulas (7) and (25) imply

$$\eta_1(z) = v.p.\int_{-\infty}^{\infty} \frac{\{sh(\pi-\alpha)\xi\, ch[\xi\Phi(z-\alpha)]-\frac{\rho}{\xi}\, ch(\pi-\alpha)\xi\,\frac{d}{dz}\, ch[\xi\Phi(z-\alpha)]\}\,\overset{+}{Q}(\xi)\,d\xi}{\overset{+}{\Gamma}(\xi;\gamma)[ch\alpha\xi\, sh(\pi-\alpha)\xi+\rho\, sh\alpha\xi\, ch(\pi-\alpha)\xi]}. \tag{33}$$

Using the representation (33), we can write

$$\eta_1(z)+\eta_1(-z)=F(-ik_1\cos z), \tag{34}$$

where

$$F(-ik_1\cos z)=\text{v.p.}\int_{-\infty}^{\infty}\frac{sh(\pi-\alpha)\xi\{ch[\xi\Phi(z-\alpha)]+ch[\xi\Phi(z+\alpha)]\}\overset{+}{Q}(\xi)d\xi}{\overset{+}{\Gamma}(\xi;\gamma)[ch\alpha\xi\, sh(\pi-\alpha)\xi+p\, sh\alpha\xi\, ch(\pi-\alpha)\xi]}-$$

$$-\text{v.p.}\int_{-\infty}^{\infty}\frac{p\, ch(\pi-\alpha)\xi\frac{d}{dz}\{ch[\xi\Phi(z-\alpha)]-ch[\xi\Phi(z+\alpha)]\}\overset{+}{Q}(\xi)d\xi}{\xi\overset{+}{\Gamma}(\xi;\gamma)[ch\alpha\xi\, sh(\pi-\alpha)\xi+p\, sh\alpha\xi\, ch(\pi-\alpha)\xi]}. \tag{35}$$

Thus the functions $\eta_1(z)$ and $\eta_2(z)$, defined by formulas (25) actually represent a solution of system (6) in the case where the outer disturbance is characterized by expression (4) in which $F(-ik_1\cos z)$ is defined for an arbitrary function $\overset{+}{Q}(\xi)$ of condition (24). It would be of greater interest, of course, to solve the problem of determining the function $\overset{+}{Q}(\xi)$ for a given function $F(-ik_1\cos z)$, which would correspond to giving an arbitrary source of initial disturbances. However, it appears to us that such a problem can be solved at the present time only approximately, and only for certain particular cases of relations between the parameters of the problem.

Substituting the value of the functions $\eta_q(z)$ into formulas (3) and interchanging the order of integration,* we obtain

$$W_1=\text{v.p.}\int_{-\infty}^{\infty}\frac{[sh(\pi-\alpha)\xi\cdot M+p\,ch(\pi-\alpha)\xi\cdot N]\overset{+}{Q}(\xi)d\xi}{\overset{+}{\Gamma}(\xi;\gamma)[ch\alpha\xi\, sh(\pi-\alpha)\xi+p\, sh\alpha\xi\, ch(\pi-\alpha)\xi]},$$

$$W_2=\text{v.p.}\int_{-\infty}^{\infty}\frac{2e^{-\frac{\pi\xi}{2}}\eta_{-i\xi}(k_2 r)\, ch(\pi-\alpha)\xi\,\overset{+}{Q}(\xi)\,d\xi}{\overset{+}{\Gamma}(\xi;\gamma)[ch\alpha\xi\, sh(\pi-\alpha)\xi+p\, sh\alpha\xi\, ch(\pi-\alpha)\xi]}, \tag{36}$$

where

$$M=\frac{1}{2\pi i}\int_{Re\,p=\sigma_0\geq 0}\frac{e^{-\xi\Phi(\varphi-\alpha+\frac{\pi}{2}-i\,arsh\frac{p}{k_1})}-e^{-\xi\Phi(\alpha-\varphi+\frac{\pi}{2}-i\,arsh\frac{p}{k_1})}}{\sqrt{p^2+k_1^2}}e^{pr}dp,$$

$$N=-\frac{1}{2\pi\xi}\int_{Re\,p=\sigma_0\geq 0}\frac{d}{dp}\left\{e^{-\xi\Phi(\varphi-\alpha+\frac{\pi}{2}-i\,arsh\frac{p}{k_1})}-e^{\xi\Phi(\alpha-\varphi+\frac{\pi}{2}-i\,arsh\frac{p}{k_1})}\right\}e^{pr}dp. \tag{37}$$

In the case $k_1=k_2=k$ equation (24) for determining the function $\overset{+}{Q}(\xi)$ degenerates into the equality

$$\overset{+}{Q}(\xi)=\overset{+}{\Psi}(\xi); \tag{38}$$

$$\overset{+}{\Gamma}(\xi;\gamma)=1. \tag{39}$$

---

*The interchange of the order of integration is justified here, since for fixed branches of the many-valued functions $\Phi(z)$ and $arsh\frac{p}{k}$ the integrals with respect to the variable ξ converge uniformly in p, and the outer integrals exist.

The solution (36), (37) hereby goes over into the exact solution

$$W_1 = v.p.\int_{-\infty}^{\infty} \frac{2[-sh(\pi-\alpha)\tau\, sh(\varphi-\alpha)\tau + p\, ch(\pi-\alpha)\tau\, ch(\varphi-\alpha)\tau]\, e^{-\frac{\pi\tau}{2}} J_{-i\tau}(\kappa\tau)\, \overset{+}{\Psi}(\tau)\, d\tau}{ch\,\alpha\tau\, sh(\pi-\alpha)\tau + p\, sh\,\alpha\tau\, ch(\pi-\alpha)\tau},$$

$$W_2 = v.p.\int_{-\infty}^{\infty} \frac{2\, ch(\pi-\varphi)\tau\, e^{-\frac{\pi\tau}{2}} J_{-i\tau}(\kappa\tau)\, \overset{+}{\Psi}(\tau)\, d\tau}{ch\,\alpha\tau\, sh(\pi-\alpha)\tau + p\, sh\,\alpha\tau\, ch(\pi-\alpha)\tau}. \tag{40}$$

The difference between formulas (36) and (40) consists, first of all, in the appearance of an additional singularity $\overset{+}{\Gamma}(\tau;\gamma)$ in the integrals with respect to the variable τ, corresponding to lateral waves, and also in the more complicated dependence on the variables τ and φ in the expression for W_1.

APPENDIX I

In the present paper the function $\Phi(z)$ is understood to mean a branch of the many-valued function $\arccos \gamma \cos z$ (the branch points $z = \pm \arccos\frac{1}{\gamma} + n\pi$, $n = 0, \pm 1, \pm 2, \ldots$ are pairwise joined by cuts) which is defined in the (z) plane by the following conditions: $\Phi(z) = z$ for $\gamma = 1$, $P(z) = \frac{d\Phi}{dz} = \frac{\gamma \sin z}{\sqrt{1-\gamma^2\cos^2 z}}$. Moreover,

$$\Phi(z+n\pi) = n\pi + \Phi(z),$$

$$P(z+n\pi) = P(z),$$

$$\Phi(-z) = -\Phi(z),$$

$$P(-z) = P(z).$$

All these relations hold if the branch in question is defined by the equations

$$\Phi(z) = n\pi + \arccos \gamma \cos(z - n\pi), \tag{41}$$

where $0 \leqslant \operatorname{Re} \arccos \gamma \cos(z - n\pi) \leqslant \pi$, $n = 0, \pm 1, \ldots$

We shall consider the proof of these relations. From (41) we have

$$\Phi(z+n\pi) = n\pi + \arccos \gamma \cos z = n\pi + \Phi(z),$$

where $0 \leqslant \operatorname{Re} \arccos \gamma \cos z \leqslant \pi$, which is equivalent to the first relation.

Relation 2 follows from relation 1 and the equation $P(z) = \frac{d\Phi}{dz}$.

To prove relation 3 we first consider the case $0 \leqslant \operatorname{Re} \arccos \gamma \cos z \leqslant \pi$. Then

$$\Phi(-z) = -\pi + \arccos \gamma \cos(\pi - z) = -\pi + \arccos(-\gamma \cos z) = -\pi + \pi - \arccos \gamma \cos z = -\arccos \gamma \cos z = -\Phi(z).$$

The result obtained is extended to arbitrary z by relation 1.

Finally, relation 4 is obtained from relation 3 and the formula $P(z) = \frac{d\Phi}{dz}$.

APPENDIX II

In this appendix formulas are presented which are of basic importance for problems with multilayer angular regions and elastic media. A representation of a certain singular integral in the form of an expansion having a singular term is also given. Since these formulas are apparently presented here for the first time, their derivation is also given.

Equations (7) and (13) imply

$$\frac{d}{dz} e^{-\Phi(z)\xi} = -\xi P(z) e^{-\Phi(z)\xi} = -\int_{-\infty}^{\infty} \xi K_1(\varkappa,\xi) e^{-z\varkappa} d\varkappa,$$
$$\frac{d}{dz} e^{-\Phi(z)\xi} = -\int_{-\infty}^{\infty} \varkappa K_2(\varkappa,\xi) e^{-z\varkappa} d\varkappa. \tag{42}$$

From this we formally obtain

$$\xi K_1(\varkappa,\xi) = -\frac{1}{2\pi i}\int_{-i\infty}^{i\infty} e^{z\varkappa}\frac{d}{dz} e^{-\Phi(z)\xi} dz = \varkappa K_2(\varkappa,\xi),$$
$$K_1(\varkappa,\xi) = \frac{\varkappa}{\xi} K_2(\varkappa,\xi). \tag{43}$$

For the chosen branch of the function $\Phi(z) = \arccos \gamma \cos z$ we have

$$K_2(-\varkappa,-\xi) = \frac{1}{2\pi i}\int_{-i\infty}^{i\infty} e^{\xi\Phi(z) - z\varkappa} dz = \frac{1}{2\pi i}\int_{-i\infty}^{i\infty} e^{-\xi\Phi(z)+z\varkappa} dz = K_2(\varkappa,\xi). \tag{44}$$

Therefore,

$$K_2^{+}(\varkappa,\xi) = \frac{K_2(\varkappa,\xi) + K_2(-\varkappa,-\xi)}{2} = \frac{1}{2\pi}\int_{-\infty}^{\infty} \cos(z\varkappa - \xi \operatorname{arccos}\gamma\, \operatorname{ch} z)\, dz . \tag{45}$$

We represent $K_2^{+}(\varkappa,\xi)$ of (45) in the form

$$K_2^{+}(\varkappa,\xi) = \frac{1}{2\pi}\int_{-\infty}^{\infty} \{\cos(\varkappa-\xi)z\, A(z) - \sin(\varkappa-\xi)z\, B(z)\}\, dz, \tag{46}$$

where

$$A(z) = \cos(z - \operatorname{arcch}\gamma\, \operatorname{ch} z)\xi,$$
$$B(z) = \sin(z - \operatorname{arcch}\gamma\, \operatorname{ch} z)\xi. \tag{47}$$

If the function $A(z)$ is transformed to the form

$$A(z) = \cos(\ln\gamma)\cos\xi y + \sin(\xi\ln\gamma)\sin\xi y,$$
$$y = \ln\left(1+\sqrt{1-\frac{1}{\operatorname{ch}^2 z}}\right) - \ln\left(1+\sqrt{1-\frac{1}{\gamma^2\operatorname{ch}^2 z}}\right) \tag{48}$$

and use is made of the expansions

$$\cos\xi y = 1 + \sum_{k=2}^{\infty}\frac{a_k(\xi)}{k!}\frac{1}{\operatorname{ch}^{2k} z},$$
$$\sin\xi y = \frac{b_1(\xi)}{\operatorname{ch}^2 z} + \sum_{k=2}^{\infty}\frac{b_k(\xi)}{k!}\frac{1}{\operatorname{ch}^{2k} z}, \tag{49}$$

in which

$$b_1(\xi)=-\frac{\xi}{4}\left(1-\frac{1}{\gamma^2}\right), \qquad a_2(\xi)=-\frac{\xi^2}{16}\left(1+\frac{1}{\gamma^2}\right)^2,$$

$$b_2(\xi)=-\frac{3\xi}{16}\left(1-\frac{1}{\gamma^4}\right), \qquad a_3(\xi)=-\frac{9\xi^2}{64}\left(1-\frac{1}{\gamma^2}\right)\left(1-\frac{1}{\gamma^4}\right), \tag{50}$$

$$b_3(\xi)=-\frac{\xi}{8}\left(1-\frac{1}{\gamma^2}\right)+\frac{\xi^3}{64}\left(1-\frac{1}{\gamma^2}\right)^3,$$

and also the equations

$$\int_0^\infty \frac{\cos\beta x}{\operatorname{ch}^{2n} x}\,dx=\frac{1}{(2n-1)!}\,\frac{\pi\beta}{2\operatorname{sh}\frac{\pi\beta}{2}}\prod_{k=1}^{n-1}(\beta^2+4k^2), \quad (n\geq 2,\ \beta>0),$$

$$\int_0^\infty \frac{\cos\beta x}{\operatorname{ch}^2 x}\,dx=\frac{\pi\beta}{2\operatorname{sh}\frac{\pi\beta}{2}}, \quad (\beta>0), \tag{51}$$

then for $\overset{+}{K}_2(\varkappa,\xi)$ we obtain the representation

$$\overset{+}{K}_2(\varkappa,\xi)=\overset{+}{\Gamma}(\xi,\gamma)\,\delta(\varkappa-\xi)+\overset{+}{R}(\varkappa,\xi;\gamma), \tag{52}$$

where $\delta(\varkappa-\xi)$ is the Dirac symbol, and

$$\overset{+}{\Gamma}(\xi;\gamma)=\cos(\xi\ln\gamma),$$

$$\overset{+}{R}(\varkappa,\xi;\gamma)=\overset{+}{R}_1(\varkappa,\xi;\gamma)+\overset{+}{R}_2(\varkappa,\xi;\gamma), \tag{53}$$

where

$$\overset{+}{R}_1(\varkappa,\xi;\gamma)=-\frac{\xi\sin(\xi\ln\gamma)}{16}\left(1-\frac{1}{\gamma^2}\right)\frac{\varkappa-\xi}{\operatorname{sh}\frac{\pi(\varkappa-\xi)}{2}}+$$

$$+\sum_{n=2}^{\infty}\frac{\cos(\xi\ln\gamma)\,a_n^+(\xi)+\sin(\xi\ln\gamma)\,b_n^-(\xi)}{4n!\,(2n-1)!}\,\frac{\varkappa-\xi}{\operatorname{sh}\frac{\pi}{2}(\varkappa-\xi)}\cdot\prod_{k=1}^{n-1}\left[(\varkappa-\xi)^2+4k^2\right],$$

$$\overset{+}{R}_2(\varkappa,\xi;\gamma)=-\frac{1}{2\pi}\int_{-\infty}^{\infty}\sin(\varkappa-\xi)z\,\sin(z-\operatorname{arcch}\gamma\operatorname{ch}z)\,\xi\,dz, \tag{54}$$

$$a_2^+(\xi)=-\frac{\xi^2}{16}\left(1-\frac{1}{\gamma^2}\right)^2, \quad a_3^+(\xi)=-\frac{9\xi^2}{64}\left(1-\frac{1}{\gamma^2}\right)\left(1-\frac{1}{\gamma^4}\right),\ \ldots$$

$$b_2^-(\xi)=-\frac{3\xi}{16}\left(1-\frac{1}{\gamma^4}\right), \quad b_3^-(\xi)=-\frac{\xi}{8}\left(1-\frac{1}{\gamma^2}\right)+\frac{\xi^3}{64}\left(1-\frac{1}{\gamma^2}\right)^3,\ \ldots$$

The author wishes to thank G. I. Petrashen' under whose supervision this work was completed and also V. M. Babich who read the manuscript and made a number of valuable suggestions.

LITERATURE CITED

1. Nikolaev, B. G., "Application of the Laplace method to the construction of solutions of the Helmholtz equation," This volume, p. 65.
2. Malyuzhenets, G. D., Doctoral Dissertation [in Russian].
3. Gradshtyn, I. S., and Ryzhik, I. M., Table of Integrals, Series, and Products [Engl. transl.], Academic Press, New York (1965).

EIGENFUNCTIONS OF THE LAPLACE OPERATOR ON THE SURFACE OF A TRIAXIAL ELLIPSOID AND IN THE REGION EXTERIOR TO IT

T. F. Pankratova

The present paper deals with the construction of asymptotic expressions for quasi-eigenvalues and -eigenfunctions of the Laplace operator in the region exterior to a triaxial ellipsoid and the eigenfunctions of the Laplace operator on the surface of the ellipsoid. Quasi-eigenvalues and -eigenfunctions and also eigenvalues and eigenfunctions mean the same as in [1] (a condensed review of these questions is found in §1 of the present paper). Quasi-eigenvalues are defined for functions concentrated at the surface of the ellipsoid. Asymptotic expressions are found for those eigenfunctions which differ appreciably from zero only in a neighborhood of the principal ellipses of the ellipsoid. The asymptotic expressions for the eigenfunctions of the Laplace operator on the surface of the ellipsoid found by the standard-equation method are compared with the expressions found by the parabolic-equation method in [2]. The complete agreement of the results confirms the conclusions of [2].

§1. Construction of the Green's Function*

We introduce an ellipsoidal system of coordinates (ξ, η, ζ) by the equation

$$\frac{x^2}{\vartheta - a^2} + \frac{y^2}{\vartheta - b^2} + \frac{z^2}{\vartheta - c^2} = 1, \ (0 < c^2 < b^2 < a^2 < \infty), \tag{1}$$

which defines the coordinate surfaces:

if $a^2 \leqslant \vartheta < \infty$ $(\vartheta \equiv \xi)$ are confocal ellipsoids;

if $b^2 \leqslant \vartheta \leqslant a^2$ $(\vartheta \equiv \eta)$ are confocal hyperboloids of a single sheet;

If $c^2 \leqslant \vartheta \leqslant b^2$ $(\vartheta \equiv \zeta)$ are confocal hyperboloids of two sheets; (x, y, and z are here the Cartesian coordinates).

The equation $\xi = \gamma$, where $\gamma = const$, defines the surface of a given ellipse.

The Green's function for the Helmholtz equation

$$(\Delta + \kappa^2) u = 0 \tag{2}$$

in the region exterior to the triaxial ellipsoid is defined as the function $G(\xi, \eta, \zeta; \xi_0, \eta_0, \zeta_0; \kappa) = G(\xi_0, \eta_0, \zeta_0; \xi, \eta, \zeta; \kappa)$ having the following properties:

*The formula for the Green's function in terms of a series of functions on the surface is well known [3]. Certain techniques used in constructing the Green's function are important in the sequel, and we therefore present this construction.

1. It satisfies the equation

$$(\Delta+\kappa^2)\, G = -4\pi\delta(M-M_0) \tag{3}$$

where $\delta(M-M_0)=\frac{1}{h_1 h_2 h_3}\delta(\xi-\xi_0)\,\delta(\eta-\eta_0)\,\delta(\zeta-\zeta_0)$ is the three-dimensional delta function; h_1, h_2, h_3 are the Lamé parameters in the ellipsoidal coordinate system.

2. The conditions

$$G\big|_{\xi=\gamma}=0 \quad \left(\text{or } \frac{\partial G}{\partial n}\Big|_{\xi=\gamma}=0\right) \tag{4}$$

are satisfied on the surface of the ellipsoid $\xi=\gamma$.

3. At infinity it satisfies the radiation condition

$$\lim_{\xi\to\infty} G=0, \qquad \operatorname{Im}\kappa>0. \tag{5}$$

It is well known [4] that the Green's function exists and is an analytic function of the parameter κ in the upper half plane; it can be continued into the lower half plane and has there no singularities except poles. The poles are called quasi-eigenvalues of the Laplace operator. The coefficients of the principal part of the Green's function at a pole satisfy conditions (2) and (4) and are called quasi-eigenfunctions.

Equation (2) admits separation of variables in ellipsoidal coordinates, that is, there are particular solutions of the form

$$u(\xi,\eta,\zeta)=X(\xi)\,Y(\eta)\,Z(\zeta). \tag{6}$$

The functions $X(\xi)$, $Y(\eta)$, $Z(\zeta)$ are solutions of the Lamé equations

$$4\sqrt{\mathcal{D}(\xi)}\frac{d}{d\xi}\left(\sqrt{\mathcal{D}(\xi)}\frac{d}{d\xi}X\right)+\kappa^2 p(\xi)X=0, \tag{7}$$

$$4\sqrt{-\mathcal{D}(\eta)}\frac{d}{d\eta}\left(\sqrt{-\mathcal{D}(\eta)}\frac{d}{d\eta}Y\right)-\kappa^2 p(\eta)Y=0, \tag{8}$$

$$4\sqrt{\mathcal{D}(\zeta)}\frac{d}{d\zeta}\left(\sqrt{\mathcal{D}(\zeta)}\frac{d}{d\zeta}Z\right)+\kappa^2 p(\zeta)Z=0, \tag{9}$$

where $\mathcal{D}(\vartheta)\equiv(\vartheta-a^2)(\vartheta-b^2)(\vartheta-c^2)$, $p(\vartheta)\equiv\vartheta^2+\alpha\vartheta+\beta=(\vartheta-\vartheta_1)(\vartheta-\vartheta_2)$, α,β are separation parameters, and ϑ_1 and ϑ_2 are roots of $p(\vartheta)$.

Equations (7), (8), and (9) are essentially the same equation considered on different segments of the ϑ axis.

The boundary conditions for the functions X, Y, Z are obtained from conditions (4) and symmetry considerations (the Cartesian coordinates (x, y, z) can be uniquely expressed in terms of the ellipsoidal coordinates (ξ,η,ζ) only in a single octant):

$$X(\xi)\big|_{\xi=\gamma}=0 \quad \text{or} \quad \frac{\partial X}{\partial \xi}\Big|_{\xi=\gamma}=0), \tag{10}$$

$$Y(\eta)\big|_{\substack{\eta=a^2\\ \eta=b^2}}=0 \quad \text{or} \quad \frac{\partial Y}{\partial \eta}\Big|_{\substack{\eta=a^2\\ \eta=b^2}}=0), \tag{11}$$

$$Z(\xi)\Big|_{\substack{\xi=c^2\\ \xi=b^2}}=0 \qquad \left(\text{или}\ \frac{\partial Z}{\partial \xi}\Big|_{\substack{\xi=c^2\\ \xi=b^2}}=0\right). \tag{12}$$

We introduce the notation

$$S(\eta,\xi)=Y(\eta)\,Z(\xi)\,, \tag{13}$$

where $Y(\eta)$ satisfies equation (8) and boundary condition (11), and $Z(\xi)$ satisfies equation (9) and condition (12). The $S(\eta,\xi)$ are the surface Lamé functions.

The numbers α and β, for which problems (8) (11) and (9), (12) are solvable are called eigenvalues of the system (8), (9) and the corresponding solutions are called eigenfunctions. However, this is not a standard problem for eigenvalues and eigenfunctions, since both eigenvalues appear in each of equations (8) and (9).

From equations (8) and (9) two different partial differential equations can be obtained which contain only one of the parameters of α or β:

$$L_\alpha S=-\alpha S\,, \tag{14}$$

$$L_\beta S=\beta S\,, \tag{15}$$

where

$$L_\alpha\equiv\frac{4}{\kappa^2(\eta-\xi)}\left[\sqrt{-D(\eta)}\,\frac{\partial}{\partial\eta}\sqrt{-D(\eta)}\,\frac{\partial}{\partial\eta}+\sqrt{D(\xi)}\,\frac{\partial}{\partial\xi}\sqrt{D(\xi)}\,\frac{\partial}{\partial\xi}\right]+\eta+\xi,$$

$$L_\beta\equiv\frac{4}{\kappa^2(\eta-\xi)}\left[\xi\sqrt{-D(\eta)}\,\frac{\partial}{\partial\eta}\sqrt{-D(\eta)}\,\frac{\partial}{\partial\eta}+\eta\sqrt{D(\xi)}\,\frac{\partial}{\partial\xi}\sqrt{D(\xi)}\,\frac{\partial}{\partial\xi}\right]+\eta\xi.$$

The functions $S(\eta,\xi)$ are eigenfunctions of the operator L_α, corresponding to eigenvalues $-\alpha$, and eigenfunctions of the operator L_β, distinct from L_α, corresponding to the eigenvalues β.

We introduce the space $L_{2,\sigma}(\Omega)$ of functions which are square-summable on the surfaces Ω of any ellipsoid with weight

$$\sigma\equiv\frac{\eta-\xi}{\sqrt{-D(\eta)D(\xi)}}\,. \tag{16}$$

The scalar product in the space $L_{2\sigma}(\Omega)$ has the form

$$(S_1,S_2)\equiv\int_{b^2}^{a^2}\int_{c^2}^{b^2}S_1(\eta,\xi)\,S_2(\eta,\xi)\frac{\eta-\xi}{\sqrt{-D(\eta)D(\xi)}}\,d\eta\,d\xi\,. \tag{17}$$

It is easy to show that for real values of the parameter κ the operators L_α and L_β are symmetric, positive, and have discrete spectra, while the functions $S(\eta,\xi)$ are orthogonal.

Suppose that to the pair α_m, β_n there corresponds $S_{mn}(\eta,\xi)$; then

$$(S_{mn},S_{ik})=\delta_{mi}\,\delta_{nk}\,E^2_{mn}\,, \tag{18}$$

where

$$E^2_{mn}=\int_{b^2}^{a^2}\int_{c^2}^{b^2}S^2_{mn}(\eta,\xi)\frac{\eta-\xi}{\sqrt{-D(\eta)D(\xi)}}\,d\eta\,d\xi,$$

and δ_{pq} is the Kronecker symbol.

Assuming that the surface Lamé functions form a complete system in $L_{2,\sigma}(\Omega)$, we construct the Green's function in the form of a Fourier series with respect to the orthonormal functions $\left\{\frac{S_{mn}}{E_{mn}}\right\}$:

$$G = \sum_{m,n} \alpha_{mn} \frac{S_{mn}}{E_{mn}}. \tag{19}$$

Using equation (3), which the Green's function satisfies, it is possible to obtain an explicity expression for the coefficients α_{mn} :

$$\alpha_{mn} \equiv \left(G, \frac{S_{mn}}{E_{mn}}\right) = -32\pi \frac{S_{mn}(\eta_0, \zeta_0, \kappa)}{E_{mn}} \frac{\mathcal{D}(\xi_0)}{\Delta_{12}} \cdot \begin{cases} X_{1mn}(\xi) X_{2mn}(\xi_0), & \xi < \xi_0, \\ X_{1mn}(\xi_0) X_{2mn}(\xi), & \xi > \xi_0, \end{cases} \tag{20}$$

where X_{1mn} and X_{2mn} are linearly independent solutions of equation (7) [X_{1mn} satisfies condition (10) on the boundary, and X_{2mn} satisfies condition (5) at infinity] and Δ_{12} is the Wronskian determinant at the point ξ_0:

$$\Delta_{12} \equiv \left[X'_{1mn}(\xi) X_{2mn}(\xi) - X_{1mn}(\xi) X'_{2mn}(\xi)\right]_{\xi=\xi_0}.$$

§2. Asymptotic Expressions for the Quasi-eigenvalues and Quasi-eigenfunctions

It is evident from formula (20) that the terms of the series (19) become infinite at points where $\Delta_{12} = 0$. It is natural to suppose that the zeros of the Wronskian determinant will be eiganvalues of the Laplace operator. The functions to which they correspond are products of solutions of the Lamé equations. The asymptotic expression for the solution of the Lamé equation depends on the distribution of the zeros ϑ_1 and ϑ_2 of the polynomial $P(\vartheta)$ in the region under consideration. In [5] and [6] the first terms are computed for the asymptotic expression for large κ of certain solutions of the Lamé equation far from the singular points under the assumption that the zeros are sufficiently far apart. In these papers the interior problem for the eigenvalues was solved. It followed from the boundary conditions that the zeros ϑ_1 and ϑ_2 of the polynomial $p(\vartheta)$ (in accordance with the usual terminology, we shall call them turning points) lie on the real axis ϑ. For the exterior problem it follows from the corresponding boundary conditions that ϑ_1 and ϑ_2 are, in general, complex.

We consider the case in which one of the turning points ϑ_1 is close to a pole of the equation, and the second ϑ_2 is close to the point γ, κ being assumed large.

(One of the turning points must be near the ray $[\gamma, \infty]$, for if this is not the case the asymptotic expression for the functions X_{jmn} $(j=1,2)$ will have the form

$$X_{1mn} = \frac{1}{\sqrt[4]{P(\xi)}} \left[e^{i\frac{\kappa}{2}\int_{\gamma}^{\xi}\sqrt{\frac{P(\xi)}{\mathcal{D}(\xi)}}d\xi} - e^{-i\frac{\kappa}{2}\int_{\gamma}^{\xi}\sqrt{\frac{P(\xi)}{\mathcal{D}(\xi)}}d\xi}\right],$$

$$X_{2mn} = \frac{1}{\sqrt[4]{P(\xi)}} e^{i\frac{\kappa}{2}\int_{\gamma}^{\xi}\sqrt{\frac{P(\xi)}{\mathcal{D}(\xi)}}d\xi},$$

$$\Delta_{12} = \frac{2i\kappa}{\sqrt{\mathcal{D}(\xi_0)}},$$

and the Wronskian determinant will have no zeros (except $\kappa = 0$)).

The standard equation for (7) near a turning point is the Airy equation

$$w'' = tw,$$

where

$$\frac{2}{3}t^{\frac{3}{2}} = i\frac{\kappa}{2}\int_{\vartheta_2}^{\xi}\sqrt{\frac{p(\xi)}{\mathfrak{D}(\xi)}}\,d\xi, \qquad t \equiv t(\xi, \kappa),$$

and the asymptotic expression for the linearly independent solutions has the form

$$X_{1mn} = \sqrt[4]{\frac{t(\xi,\kappa)t(\gamma,\kappa)}{p(\xi)p(\gamma)}}\left\{W_1[t(\gamma,\kappa)]W_2[t(\xi,\kappa)] - W_1[t(\xi,\kappa)]W_2[t(\gamma,\kappa)]\right\},$$

$$X_{2mn} = \sqrt[4]{\frac{t(\xi,\kappa)}{p(\xi)}}\,W_2[t(\xi,\kappa)],$$

where

$$W_1 = \frac{1}{\sqrt{\pi}}\int_{\infty\exp(\frac{4\pi i}{3})}^{\infty} e^{tz - \frac{z^3}{3}}\,dz,$$

$$W_2 = \frac{1}{\sqrt{\pi}}\int_{\infty\exp(\frac{2\pi i}{3})}^{\infty} e^{tz - \frac{z^3}{3}}\,dz.$$

The Wronskian determinant Δ_{12} for these solutions is given by the expression

$$\Delta_{12} = W_2[t(\gamma,\kappa)]\frac{2i\kappa}{\sqrt{\mathfrak{D}(\xi_0)}},$$

and becomes zero at those values of κ for which

$$W_2[t(\gamma,\kappa)] = 0.$$

For $\vartheta_2 \simeq \gamma$ we have

$$\int_{\vartheta_2}^{\gamma}\sqrt{\frac{p(\xi)}{\mathfrak{D}(\xi)}}\,d\xi \simeq -\sqrt{\frac{\gamma-\vartheta_1}{\mathfrak{D}(\gamma)}}\,\frac{2}{3}(\gamma-\vartheta_2)^{\frac{3}{2}}$$

and the quasi-eigenvalues have the form

$$\kappa_{mns} = -\frac{2}{i}\sqrt{\frac{\mathfrak{D}(\gamma)}{\gamma-\vartheta_{1m}}}\left(\frac{t_s}{\gamma-\vartheta_{2n}}\right)^{\frac{3}{2}}, \tag{23}$$

where t_s is a root of the Airy function $W_2(t)$ $(s=1,2,\ldots)$, and ϑ_1 and ϑ_2 are related to the eigenvalues α, β and are found by solving problems (8), (11) and (9), (12).

In the expression (6) for the eigenfunctions, the factor depending on the coordinates ξ has the form

$$X(\xi) = \text{const}\; W\left[\left(\frac{3i\kappa}{4}\right)^{\frac{2}{3}}(\xi-\gamma)\right]. \tag{24}$$

For values of ϑ_2, near γ the eigenfunctions concentrate at the surface of the ellipsoid.

We consider now equation (8), (9). Let

$$\vartheta_1 = c^2 + \kappa^{-1}\mu, \qquad \mu = O(1). \tag{25}$$

By a change of variables

$$c^{\frac{1}{2}}\kappa^{-1}t^2 = \xi - c^2,$$

where

$$C \equiv \frac{\vartheta_2 - c^2}{(a^2-c^2)(b^2-c^2)},$$

equation (9) is easily reduced [neglecting terms of order $O(\frac{1}{\kappa})$] to the well-known equation

$$\frac{d^2 z}{dt^2} + (\lambda - t^2)z = 0,$$

$$\lambda = \mu C^{\frac{1}{2}}.$$

This equation has solutions which decrease at infinity for $\lambda = 2m+1$.

The first term of the asymptotic expression is

$$z(\xi) = e^{-\frac{1}{2}C^{\frac{1}{2}}\kappa(\xi - c^2)} \cdot H_m[C^{\frac{1}{4}}\kappa^{\frac{1}{2}}(\xi - c^2)^{\frac{1}{2}}] \tag{26}$$

with

$$\vartheta_{1m} = c^2 + \kappa^{-1}C^{-\frac{1}{2}}(2m+1), \qquad m = O(1). \tag{27}$$

$H_m(z)$ is here a Hermite function.

Equation (8) under condition (25) has the asymptotic solution

$$Y(\eta) = \frac{1}{\sqrt[4]{P(\eta)}} e^{\pm i\frac{\kappa}{2}\int_{b^2}^{\eta}\sqrt{\frac{\eta-\vartheta_2}{(\eta-a^2)(\eta-b^2)}}\,d\eta - i(m+\frac{1}{2})C^{\frac{1}{2}}\frac{1}{2}\int_{b^2}^{\eta}\frac{1}{\eta-c^2}\sqrt{\frac{\eta-\vartheta_2}{(\eta-a^2)(\eta-b^2)}}\,d\eta}. \tag{28}$$

The boundary conditions (10) give

$$\kappa\int_{b^2}^{\eta}\sqrt{\frac{\eta-\vartheta_{2n}}{(\eta-a^2)(\eta-b^2)}}\,d\eta - (m+\frac{1}{2})C^{\frac{1}{2}}\int_{b^2}^{\eta}\frac{1}{\eta-c^2}\sqrt{\frac{\eta-\vartheta_{2n}}{(\eta-a^2)(\eta-b^2)}}\,d\eta = n\pi. \tag{29}$$

Equations (23), (27), and (29) define quasi-eigenvalues κ_{mns} for functions concentrated in a neighborhood of the "minimal" ellipse of the ellipsoid

$$\frac{x^2}{\gamma - a^2} + \frac{y^2}{\gamma - b^2} = 1, \quad z = 0,$$

the asymptotic expression for which is given by formulas (6), (24), (26), and (28).

Asymptotic expressions for functions concentrated near the "maximal" ellipse of the ellipsoid

$$\frac{y^2}{\gamma - b^2} + \frac{z^2}{\gamma - c^2} = 1, \qquad x = 0,$$

are found in a similar manner. Under the condition

$$\vartheta_1 = a^2 - \kappa^{-1}\mu', \qquad \mu' = O(1),$$

the following expressions are obtained:

$$Y(\eta) = e^{-\frac{1}{2}C_1(a^2-\eta)\kappa} H_m[C_1^{\frac{1}{4}}\kappa^{\frac{1}{2}}(a^2-\eta)^{\frac{1}{2}}],$$

$$C_1 = \frac{\vartheta_2 - a^2}{(c^2-a^2)(b^2-a^2)}, \tag{30}$$

$$\vartheta_{1m} = a^2 - \kappa^{-1} c_1 (2m+1), \qquad m = O(1), \tag{31}$$

$$z(\xi) = \frac{1}{\sqrt[4]{p(\xi)}} e^{\pm i \frac{\kappa}{2} \int_{c^2}^{\xi} \sqrt{\frac{\xi - \vartheta_{2n}}{(\xi - c^2)(\xi - b^2)}}\, d\xi - i\left(m+\frac{1}{2}\right) C_1^{\frac{1}{2}} \int_{c^2}^{\xi} \frac{1}{\xi - a^2} \frac{1}{2} \sqrt{\frac{\xi - \vartheta_{2n}}{(\xi - c^2)(\xi - b^2)}}\, d\xi}, \tag{32}$$

$$\kappa \int_{c^2}^{\xi} \sqrt{\frac{\xi - \vartheta_{2n}}{(\xi - c^2)(\xi - b^2)}}\, d\xi - \left(m+\frac{1}{2}\right) C_1^{\frac{1}{2}} \int_{c^2}^{\xi} \frac{1}{\xi - a^2} \sqrt{\frac{\xi - \vartheta_{2n}}{(\xi - c^2)(\xi - b^2)}}\, d\xi = n\pi. \tag{33}$$

Formulas (6), (24), (30), and (32) give the quasi-eigenfunctions, and (23), (31), and (33) give the corresponding quasi-eigenvalues (in the asymptotic sense).

The asymptotic expressions for the quasi-eigenfunctions concentrated near the intermediate ellipse

$$\frac{x^2}{\gamma - a^2} + \frac{y^2}{\gamma - b^2} = 1, \qquad y = 0,$$

on the ellipsoid cannot be obtained, that is, the hypothesis $\vartheta_1 = b^2 \pm \kappa^{-1} \mu''$, $\mu'' = O(1)$, leads to unbounded solutions on the interval (b^2, ϑ_1). Asymptotic expressions for the Lamé functions on the interval (c^2, ϑ_1) for $\vartheta_1 = b^2 - \kappa^{-1} \mu''$, or on (ϑ_1, a^2), for $\vartheta_1 = b^2 + \kappa^{-1} \mu''$, may be found in [5, 6]. Eigenfunctions are obtained which are different from zero in a wide neighborhood of the minimal or maximal ellipses.

The minimal and maximal ellipses are stable geodesics on the ellipsoid, while the intermediate ellipse is an unstable geodesic. The results obtained make it possible to conclude that the eigenfunctions are concentrated in a neighborhood of the stable closed geodesics.

§3. Eigenfunctions of the Laplace Operator on the Surface of the Ellipsoid

We now consider the Laplace operator on the surface of the ellipsoid, that is, the Beltrami operator

$$\Delta_2 = \frac{4}{\xi - \eta} \sqrt{\frac{\mathcal{D}(\eta)\mathcal{D}(\xi)}{(\eta - \gamma)(\xi - \gamma)}} \frac{\partial}{\partial \eta}\left(\frac{\xi - \gamma}{\eta - \gamma} \sqrt{\frac{\mathcal{D}(\eta)}{\mathcal{D}(\xi)}} \frac{\partial}{\partial \eta}\right) + \frac{\partial}{\partial \xi}\left(\frac{\eta - \gamma}{\xi - \gamma} \sqrt{\frac{\mathcal{D}(\xi)}{\mathcal{D}(\eta)}} \frac{\partial}{\partial \xi}\right). \tag{34}$$

The equation $(\Delta_2 + \kappa)\tilde{u} = 0$ admits separation of variables, that is, there is a particular solution of the form

$$\tilde{u}(\eta, \xi) = \tilde{Y}(\eta) \tilde{Z}(\xi), \tag{35}$$

where $\tilde{Y}(\eta)$ and $\tilde{Z}(\xi)$ are solutions of the equations

$$4\sqrt{\frac{\mathcal{D}(\eta)}{\eta - \gamma}} \frac{d}{d\eta}\left(\sqrt{\frac{\mathcal{D}(\eta)}{\eta - \gamma}} \frac{d}{d\eta} \tilde{Y}(\eta)\right) + \kappa^2 (\eta - \tilde{\vartheta}_1) \tilde{Y}(\eta) = 0, \tag{36}$$

$$4\sqrt{\frac{\mathcal{D}(\xi)}{\xi - \gamma}} \frac{d}{d\xi}\left(\sqrt{\frac{\mathcal{D}(\xi)}{\xi - \gamma}} \frac{d}{d\xi} \tilde{Z}(\xi)\right) - \kappa^2 (\xi - \tilde{\vartheta}_1) \tilde{Z}(\xi) = 0, \tag{37}$$

and $\tilde{\vartheta}_1$ is a separation parameter.

The eigenfunctions of the Beltrami operator must be single-valued on the surface of the ellipsoid. This requirement gives boundary conditions for equations (36) and (37) similar to (11) and (12).

The operator Δ_2 differs from both L_α and L_β [cf. §1, formulas (14) and (15)], and the eigenfunctions of the Beltrami operator are not equal to the surface Lamé functions. As an example, we compare the asymptotic expressions for functions concentrated near the minimal ellipse on the ellipsoid.

Let

$$\tilde{\vartheta}_1 = c^2 + \kappa^{-1}\mu, \qquad \mu = O(1). \tag{38}$$

The appropriate change of variables (see § 2) reduces equation (37) to the equation for functions of the parabolic cylinder. The first term in the asymptotic expression for $\tilde{Z}(\varepsilon)$ is given by the formula

$$\tilde{Z}(\varepsilon) = Z(\varepsilon) \qquad (\text{for } \tilde{\vartheta}_1 = \vartheta_1 \quad) \tag{39}$$

Equation (36) under condition (38) has the asymptotic solution

$$\tilde{Y}(\eta) = \frac{1}{\sqrt[4]{\eta - \tilde{\vartheta}_1}}\, e^{\pm i\frac{\kappa}{2}\int_{b^2}^{\eta}\sqrt{\frac{\eta-\gamma}{(\eta-a^2)(\eta-b^2)}}\,d\eta - i\left(m+\frac{1}{2}\right)c^{\frac{1}{2}}\int_{b^2}^{\eta}\frac{1}{\eta-c^2}\frac{1}{2}\sqrt{\frac{\eta-\gamma}{(\eta-a^2)(\eta-b^2)}}\,d\eta}. \tag{40}$$

Equalities (35), (39), and (40) give the formula

$$\tilde{U}(\eta,\varepsilon) = \frac{1}{\sqrt[4]{\eta-\tilde{\vartheta}_1}} \cdot e^{\pm i\frac{\kappa}{2}\int_{b^2}^{\eta}\sqrt{\frac{\eta-\gamma}{(\eta-a^2)(\eta-b^2)}}\,d\eta - i\left(m+\frac{1}{2}\right)c^{\frac{1}{2}}\int_{b^2}^{\eta}\frac{1}{\eta-c^2}\frac{1}{2}\sqrt{\frac{\eta-\gamma}{(\eta-a^2)(\eta-b^2)}}\,d\eta - \frac{1}{2}c^{\frac{1}{2}}\kappa(\varepsilon-c^2)} \cdot H_m\left[c^{\frac{1}{4}}\kappa^{\frac{1}{2}}(\varepsilon-c^2)^{\frac{1}{2}}\right]. \tag{41}$$

Comparing (26), (28), and (41), it is evident that for $\vartheta_2 \cong \gamma$

$$S(\eta,\varepsilon) = \frac{1}{\sqrt[4]{\eta-\gamma}}\,\tilde{U}(\eta,\varepsilon). \tag{42}$$

It is interesting to note that

$$\frac{1}{\sqrt[4]{\eta-\gamma}} = \frac{const}{\rho^{\frac{1}{6}}}, \tag{43}$$

ρ is the radius of curvature of the minimal ellipse).

We now compare formula (41) with the expression

$$\hat{U}(s,\xi) = \frac{1}{\sqrt{j(s)}}\, e^{i\kappa s - i\int_0^s \frac{m+\frac{1}{2}}{j^2(s)}\,ds + i\kappa\frac{j'}{2j}\xi^2 - \frac{\kappa}{2}\frac{\xi^2}{j^2}} \cdot H_m\left(\kappa^{\frac{1}{2}}\frac{\xi}{j(s)}\right), \tag{44}$$

obtained in [2] by the parabolic-equation method. In formula (44) the notation is as follows: (s, ξ) are semigeodesic coordinates, s is the variable of length of the geodesic, ξ is the distance to the geodesic along the surface of the ellipsoid, $j(s)$ is a solution of the equation

$$j'' + K(s)\,j = \frac{1}{j^3}, \tag{45}$$

where $K(s)$ is the Gaussian curvature of the surface on the geodesic.

We express the coordinates (η,ε) in terms of (s,ξ) with an accuracy up to terms of order ξ^2. We then obtain

$$\varepsilon - c^2 = \frac{\xi^2}{C(\eta-c^2)}, \tag{46}$$

$$\frac{1}{2}\int_{b^2}^{\eta}\sqrt{\frac{\eta-\gamma}{(\eta-b^2)(\eta-a^2)}}\,d\eta = s + \frac{j'}{2j}\xi^2 \tag{47}$$

$$j(\eta) = c^{\frac{1}{4}}\sqrt{\eta-c^2}, \tag{48}$$

where $j(\eta)$ is a solution of equation (45) in the coordinates η.

Considering (46), (47), and (48), equation (41) goes over into equation (44) obtained by the parabolic-equation method.

The author would like to thank V. M. Babich for his assistance in the work on this paper.

LITERATURE CITED

1. Babich, V. M., "The asymptotic behavior of "quasi-eigenvalues" of the exterior problem for the Laplace operator," in: Topics in Mathematical Physics, Vol. 2, M. Sh. Birman, ed., Consultants Bureau, New York (1968).
2. Babich, V. M., and Lazutkin, V. F., "Eigenfunctions concentrated near a closed geodesic," in: Topics in Mathematical Physics, Vol. 2, M. Sh. Birman, ed., Consultants Bureau, New York (1968).
3. Morse, P. M., and Feshbach, H., Methods of Theoretical Physics, McGraw-Hill, New York (1953).
4. Marchenko, V. A., and Khruslov, E. Ya., "Analytic properties of the resolvent of a certain boundary value problem," Third All-Union Symposium on Wave Diffraction, Tblisi, 24-30 September, 1964, Report [in Russian], Nauka.
5. Bykov, V. P., "The geometric optics of three-dimensional oscillations in open resonators," in: High-Power Electronics [in Russian], Vol. 4, Moscow (1965).
6. Vainshtein, L. A., "Ray flows on the triaxial ellipsoid," in: High-Power Electronics [in Russian], Vol. 4, Moscow (1965).

CALCULATION OF THE WAVE FIELDS FOR MULTIPLE WAVES NEAR THE POINTS OF ORIGIN

N. S. Smirnova

In the present paper we obtain computational formulas for the principal parts of displacement fields of multiple reflected and head waves propagating in elastic media near the point of origin of the head wave. Similar formulas for simpler cases have been considered in [1, 2].

§1. Formulation of the Problem and Certain Questions of Analysis

1. Suppose that in a system of cylindrical coordinates (r, θ, z) plane-parallel elastic layers $(0<z<z_1=h_1,\ z_1<z<z_2=h_1+h_2, \ldots, z_{n-1}<z<z_n=h_1+\ldots+h_n)$ and a half space $z>z_n=h_1+\ldots+h_n$ are given (Fig. 1).

The elastic layer numbered κ is characterized by a thickness h_κ and constant values of the density ρ_κ and speeds v_p^κ and v_s^κ of the longitudinal and transverse waves. On the boundaries separating the elastic layers we assume that the components (q, p, w) of the displacement vector $\vec{u}$ and the components t_{rz}, $t_{\theta z}$ and t_{zz} of the stress tensor are continuous. On the free surface $z=0$ the stresses are either zero or take values corresponding to some chosen boundary action. We further assume that at time $t<0$ the medium is at rest and at time $t=0$ an axially symmetric source of vibrations is switched on; the source is located either on the free surface or at any interior point of the medium. The time-dependence of the source is characterized by the function $f(t)$. Waves of all possible types begin to propagate in the elastic system. The problem is to study the reflected and head waves formed in the n-th horizon of the medium.

2. It is known [3] that the wave field caused by a wave of arbitrary type at points of the free surface can be represented in the form of Fourier-Bessel integrals

$$q(r,t)=\int_0^\infty J_1(\kappa r)\, S(\kappa,t)\, d\kappa,$$
$$w(r,t)=\int_0^\infty J_0(\kappa r)\, R(\kappa,t)\, d\kappa, \tag{1.1}$$

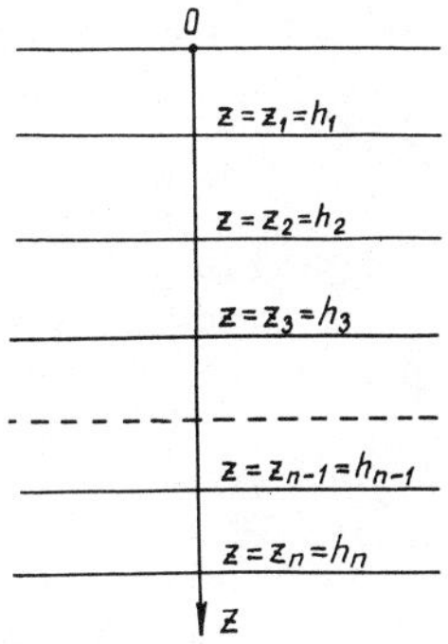

Fig. 1. (Note: the correspondence between the z_κ and h_κ in the figure is not the same as that in the text – Trans.)

Mellin integrals

$$S(\kappa,t)=\frac{1}{2\pi i}\int_{\sigma-i\infty}^{\sigma+i\infty} U(\eta)\, e^{\kappa\psi(\eta,t)}\, d\eta,$$
$$R(\kappa,t)=\frac{1}{2\pi i}\int_{\sigma-i\infty}^{\sigma+i\infty} V(\eta)\, e^{\kappa\psi(\eta,t)}\, d\eta, \tag{1.2}$$

where $\sigma > 0$ and $J_\nu(\kappa r)(\nu = 0, 1)$ are Bessel functions. (Since the problem in question is axially symmetric, it follows that $p = 0$.) In the formulas (1.2) the phase function $\varphi(\eta, t)$ has the form

$$\varphi(\eta, t) = v_s' t \eta - \sum_{\kappa=1}^{n} h_\kappa (m_\kappa \alpha_\kappa + n_\kappa \beta_\kappa), \tag{1.3}$$

where m_κ and n_κ are the number of times the disturbance passes through the κ-th layer in the form of a longitudinal and transverse wave, and α_κ and β_κ are given by

$$\alpha_\kappa = \sqrt{1 + \gamma_\kappa^2 \eta^2}, \quad \beta_\kappa = \sqrt{1 + \delta_\kappa^2 \eta^2}, \quad \gamma_\kappa = \frac{v_s'}{v_p^\kappa}, \quad \delta_\kappa = \frac{v_s'}{v_s^\kappa},$$

with $\arg \alpha_\kappa = \arg \beta_\kappa = 0$ for $\eta > 0$. The functions $U(\eta)$ and $V(\eta)$ of (1.2) are represented as products of functions $U_\kappa(\eta)$ and $V_\kappa(\eta)$

$$U(\eta) = \prod_{\kappa=1}^{n} U_\kappa(\eta), \qquad V(\eta) = \prod_{\kappa=1}^{n} V_\kappa(\eta), \tag{1.4}$$

which depend on the properties of adjacent layers and are rational functions of radicals.

3. It is well known that the principal part of the wave field is obtained by evaluating the integrals (1.2) by the saddle-point method (or the method of stationary phase) [4]. A fundamental rol͡e in this method is played by the saddle points η_0 of the phase function which satisfy the equation $\varphi'(\eta_0, t) = 0$ and are distributed on the imaginary axis symmetrically with respect to the real axis.

To each saddle point there corresponds a wave beam in the elastic medium; the quantity $|\eta_0|$ is hereby inversely proportional to the sine of the angle of incidence of the beam on the boundary separating the media. Therefore, for small angles of incidence the saddle points are located at an extremely large distance from the origin of the coordinate system, but as the angle of incidence increases they monotonically approach the branch point of maximum modulus of the phase function and can be arranged at the origin near certain branch points of the functions $U(\eta)$ and $V(\eta)$. The principal parts of the Mellin integral are obtained by evaluating integrals (1.2) over the principal segments $(\ell_\varepsilon)_\pm$ of the stationary contours passing through the saddle points and also over the principal segements $(\lambda_\varepsilon)_\pm$ of the loops enclosing the cuts going from the branch points $\eta = \pm \frac{i}{\gamma_\kappa}$ and $\eta = \pm \frac{i}{\delta_\kappa}$ in the left half plane to $-\infty$, parallel to the real axis. (Here the + sign denotes a neighborhood of a saddle point in the upper half plane, while the − sign denotes a neighborhood in the lower half plane of the variable η. In the sequel we shall restrict our consideration to the upper half plane $\operatorname{Im} \eta > 0$.) The principal parts computed along the contours $(\ell_\varepsilon)_\pm$ determine the field of the reflected wave, and those computed along the contours $(\lambda_\varepsilon)_\pm$ determine the field of the head wave, where 2ε is the length of the path of integration. Certain conditions, which are called conditions for the applicability of the method of stationary phase, must be satisfied in order to carry out the calculations indicated. If they are fulfilled, then the computations lead to the correct result for the wave fields. If these conditions are not satisfied, then the result may differ even qualitatively from the correct result.

Usually the applicability condition is not met when the saddle point is located near singularities of the functions appearing in the integrands of (1.3) and (1.4). In such cases the method of stationary phase must be applied in various modifications which are chosen such that the corresponding applicability conditions are satisfied. The papers [1, 2, 5] consider various modifications of this method. The present paper is a natural continuation of them, inasmuch as an attempt is here made to apply the variations of the method already available to obtaining the principal parts of the wave fields of multiple waves of arbitrary type.

§2. Derivation of Formulas for the Principal Parts of the Wave Fields of Multiple Waves

1. We consider the elastic medium mentioned in subsection 1 of Section 1 and assume that

$$\delta_{n+1} > \max(\gamma_\kappa, \delta_\kappa), \qquad \kappa = 1, \ldots n. \tag{2.1}$$

(It will subsequently be possible to avoid this assumption.) We shall also assume that there are no boundaries with low speed jumps in the elastic system.

We consider an arbitrary longitudinal wave reflected q times from the last n-th horizon of the system and an observer located near the point of origin of a head wave of longitudinal type appearing at points of the n-th horizon. The distribution of singularities in the complex η plane are as follows. Because of inequality (2.1), there is one branch point $\eta=\frac{i}{\gamma_{n+1}}$ of the functions $U(\eta)$ and $V(\eta)$, located on the imaginary axis above the branch point $\eta=\frac{i}{\gamma_n}$ of maximum modulus of the phase function. The saddle point η_0 is located near the point $\eta=\frac{i}{\gamma_{n+1}}$ and may be either above it or below it. In the first case angles of incidence up to the limiting angle are considered, and in the second case those past the limiting angle. In the region of angles up to the limiting angle we represent the integral $R(\kappa,t)$ (or $S(\kappa,\eta)$) as a sum of two integrals $R=\mathcal{J}_1+\mathcal{J}_3$, the first of which extends over the principal segment of the path of integration $(\ell_\varepsilon)_+$, and the second over the remaining part $(\ell)-(\ell_\varepsilon)_+$. In the region of angles past the limiting angle the same integral is represented in the form

$$R=\mathcal{J}_1+\mathcal{J}_2+\bar{\mathcal{J}}_3,$$

where $\mathcal{J}_1$ and $\bar{\mathcal{J}}_3$ are as before, and $\mathcal{J}_2$ is an integral over the principal segment of the contour $(\lambda_\varepsilon)_+$, enclosing the cut. The applicability conditions of the method, which we assume satisfied, guarantee that $\mathcal{J}_3$ and $\bar{\mathcal{J}}_3$, are small, such that

$$R=\begin{cases}\mathcal{J}_1, & \text{up to the limiting angle;}\\ \mathcal{J}_1+\mathcal{J}_2, & \text{past the limiting angle.}\end{cases}$$

We proceed to the computation of $\mathcal{J}_1$. The function $V(\eta)$ (or $U(\eta)$) of (1.2) can be represented in the form

$$V(\eta)=\bar{V}(\eta)P^q_{nn}(n,n+1)=\bar{V}(\eta)\left[A^{(0)}(\eta)+\alpha_{n+1}A^{(1)}(\eta)\right]^q, \tag{2.2}$$

where $P_{nn}(n,n+1)$ is a complex function which is the analogue* of the reflection coefficient of the longitudinal wave from the n-th boundary of separation, $A^{(0)}(\eta)$ is the part of the coefficient not containing the branch $\eta=\frac{i}{\gamma_{n+1}}$ (or the radical α_{n+1}), and $A^{(1)}(\eta)$ is the factor going with α_{n+1} in this coefficient. In order to calculate $R(\kappa,t)$ we approximate the functions $V(\eta)$ and $\varphi(\eta)$ on the principal segment $(\ell_\varepsilon)_+$ of the path of integration by the following expressions:

$$\begin{aligned} V(\eta)&\approx\bar{V}(\eta_0)\left[A^{(0)}(\eta_0)+\alpha_{n+1}A^{(1)}(\eta_0)\right]^q,\\ \varphi(\eta)&\approx\varphi(\eta_0)+\frac{\varphi''(\eta_0)}{2!}(\eta-\eta_0)^2,\\ \alpha_{n+1}&\approx e^{i\frac{\pi}{4}}\left(\eta-\frac{i}{\gamma_{n+1}}\right)^{\frac{1}{2}}\sqrt{\gamma_{n+1}^2|\eta_0|+\gamma_{n+1}}\,. \end{aligned} \tag{2.3}$$

*By this we understand a complex function which is a factor of the function $V(\eta)$ (or $U(\eta)$) and depends on the properties of the adjacent layers from the boundary of which reflection occurs. Moreover, if the value at the saddle point $\eta=i|\eta_0|$ is substituted for η in this function, then the function in question becomes the usual coefficient of reflection of a plane wave from the boundary of two elastic media. Complex analogues of the coefficients of refraction, conversion, and directedness of the source are similar expressions which we shall use in the sequel.

In the last expression

$$\arg\left(\eta-\frac{i}{\gamma_{n+1}}\right)^{\frac{1}{2}}=-\frac{\pi}{4}\,, \text{ for } \eta=\eta_0 \text{ and } |\eta_0|<\frac{1}{\gamma_{n+1}}\,, \tag{2.4}$$

and

$$\arg\left(\eta-\frac{i}{\gamma_{n+1}}\right)^{\frac{1}{2}}=\frac{\pi}{4}\,, \quad \text{for } \eta=\eta_0 \text{ and } |\eta_0|>\frac{1}{\gamma_{n+1}}\,.$$

The Mellin integral over the principal part $(\ell_\varepsilon)_+$ of the path of integration is given by the expression

$$\mathcal{J}_1=\frac{1}{2\pi i}\int\limits_{(\ell_\varepsilon)_+}\bar V(\eta_0)\left[A^{(0)}(\eta_0)+e^{i\frac{\pi}{4}}\sqrt{\gamma_{n+1}^2|\eta_0|+\gamma_{n+1}}\left(\eta-\frac{i}{\gamma_{n+1}}\right)^{\frac{1}{2}}A^{(1)}(\eta_0)\right]^q\cdot e^{i\kappa|\varphi(\eta_0)|-\frac{\kappa|\varphi''(\eta_0)|}{2!}(\eta-\eta_0)^2}\,d\eta\,. \tag{2.5}$$

Putting

$$\eta-\eta_0=\rho e^{i\frac{3}{4}\pi}\,, \tag{2.6}$$

we obtain

$$\mathcal{J}_1=\frac{1}{2\pi}e^{i(\kappa|\varphi(\eta_0)|+\frac{\pi}{4})}\bar V(\eta_0)\int\limits_{-\varepsilon}^{\varepsilon}\left[A^{(0)}(\eta_0)+e^{\frac{\pi}{4}i}\sqrt{\gamma_{n+1}^2|\eta_0|+\gamma_{n+1}}\cdot e^{i\frac{3}{8}\pi}(\rho-\beta)^{\frac{1}{2}}\cdot A^{(1)}(\eta_0)\right]^q e^{-\frac{\kappa|\varphi''(\eta_0)|}{2}\rho^2}\,d\rho\,. \tag{2.7}$$

Replacing in (2.7) the value ε by ∞, which is possible since the error hereby made is negligible, and going over to the new variable

$$y=\rho\sqrt{\frac{\kappa|\varphi''(\eta_0)|}{2}}\,, \tag{2.8}$$

we have

$$\mathcal{J}_1=\frac{1}{2\pi}e^{i(\kappa|\varphi(\eta_0)|+\frac{\pi}{4})}\bar V(\eta_0)\sqrt{\frac{2}{\kappa|\varphi''(\eta_0)|}}\int\limits_{-\infty}^{\infty}\left[A^{(0)}(\eta_0)+e^{i\frac{5}{8}\pi}\left(\frac{2}{\kappa|\varphi''(\eta_0)|}\right)^{\frac{1}{4}}\cdot\sqrt{\gamma_{n+1}^2|\eta_0|+\gamma_{n+1}}\cdot\sqrt{y-a_1}\,A^{(1)}(\eta_0)\right]^q e^{-y^2}\,dy=\frac{1}{2\pi}\sqrt{\frac{2}{\kappa|\varphi''(\eta_0)|}}\,\bar V(\eta_0)\cdot$$

$$\cdot e^{i(\kappa|\varphi(\eta_0)|+\frac{\pi}{4})}I_{\text{refl}}(a_1)=\frac{\exp[\kappa|\varphi(\eta_0)|+\frac{\pi}{4}]i}{\sqrt{2\pi\kappa|\varphi''(\eta_0)|}}\,\mathcal{J}_w^{\text{refl}}(\kappa,\eta_0)\,, \tag{2.9}$$

where

$$a_1=\left(|\eta_0|-\frac{i}{\gamma_{n+1}}\right)e^{i\frac{3}{4}\pi}\left[\frac{\kappa|\varphi''(\eta_0)|}{2}\right]^{\frac{1}{2}}\,,$$

$$\mathcal{J}_w^{\text{refl}}=\frac{1}{\sqrt{\pi}}\bar V(\eta_0)I_{\text{refl}}(a_1)\,, \tag{2.10}$$

and $\arg\sqrt{y-a_1}=-\frac{\pi}{8}$ for $y=0$ in the case of observers up to the point of origin of the head wave, and $\arg\sqrt{y-a_1}=-\frac{5}{8}\pi$ for $y=0$ in the case of observers past the point of origin of the head wave.

2. We consider the approximate evaluation of the integral over the principal segment $(\lambda_\varepsilon)_+$ of the sides of the cut joined to the vertex $\eta=\frac{i}{\gamma_{n+1}}$. The cut goes along the stationary contour of the phase

function defined by the equation $\operatorname{Im}\varphi(\eta)=\operatorname{Im}\varphi(\frac{i}{\gamma_{n+1}})$. We shall assume that the following approximate representations are valid on the principal segment of the path of integration:

$$V(\eta)\approx\bar{V}(\frac{i}{\gamma_{n+1}})[A^{(0)}(\frac{i}{\gamma_{n+1}})+e^{\frac{\pi}{4}i}\sqrt{2\gamma_{n+1}}\sqrt{\eta-\frac{i}{\gamma_{n+1}}}A^{(1)}(\frac{i}{\gamma_{n+1}})]^q,$$

$$\varphi(\eta)\approx\varphi(\frac{i}{\gamma_{n+1}})+\varphi'(\frac{i}{\gamma_{n+1}})(\eta-\frac{i}{\gamma_{n+1}})+\varphi''(\frac{i}{\gamma_{n+1}})\frac{1}{2!}(\eta-\frac{i}{\gamma_{n+1}})^2, \tag{2.11}$$

$$\text{and}\quad \arg\sqrt{\eta-\frac{i}{\gamma_{n+1}}}=\frac{\pi}{4},\qquad \text{for}\quad \eta=\eta_0 \text{ and } |\eta_0|>\frac{1}{\gamma_{n+1}}$$

We substitute (2.11) into the Mellin integral (1.2) and obtain

$$\mathcal{Y}_2=\frac{1}{2\pi i}\int_{(\lambda_\varepsilon)_+}\bar{V}(\frac{i}{\gamma_{n+1}})e^{i\kappa|\varphi(\frac{i}{\gamma_{n+1}})|}[A^{(0)}(\frac{i}{\gamma_{n+1}})+e^{\frac{\pi}{4}i}\sqrt{2\gamma_{n+1}}\sqrt{\eta-\frac{i}{\gamma_{n+1}}}A^{(1)}(\frac{i}{\gamma_{n+1}})]^q\cdot\exp[\varphi'(\frac{i}{\gamma_{n+1}})(\eta-\frac{i}{\gamma_{n+1}})+\frac{\varphi''(\frac{i}{\gamma_{n+1}})}{2!}(\eta-\frac{i}{\gamma_{n+1}})^2]d\eta, \tag{2.12}$$

or, introducing the new variable

$$t=(\eta-\frac{i}{\gamma_{n+1}})\sqrt{\frac{\kappa|\varphi''(\frac{i}{\gamma_{n+1}})|}{2!}}\,e^{-\frac{3}{4}\pi i}, \tag{2.13}$$

we rewrite (2.12) in the form

$$\mathcal{Y}_2=\frac{1}{2\pi i}e^{i[\kappa|\varphi(\frac{i}{\gamma_{n+1}})|+\frac{3}{4}\pi]}\sqrt{\frac{2}{\kappa|\varphi''(\frac{i}{\gamma_{n+1}})|}}\bar{V}(\frac{i}{\gamma_{n+1}})\cdot\Big\{-\int_0^\infty[A^{(0)}(\frac{i}{\gamma_{n+1}})-e^{\frac{5}{4}\pi i}\sqrt{2\gamma_{n+1}}(\frac{2}{\kappa|\varphi''(\frac{i}{\gamma_{n+1}})|})^{\frac{1}{4}}A^{(1)}(\frac{i}{\gamma_{n+1}})\sqrt{t}]^q\cdot$$

$$\cdot e^{-t^2-2at}dt+\int_0^\infty[A^{(0)}(\frac{i}{\gamma_{n+1}})+e^{\frac{5}{4}\pi i}\sqrt{2\gamma_{n+1}}(\frac{2}{\kappa|\varphi''(\frac{i}{\gamma_{n+1}})|})^{\frac{1}{4}}\cdot A^{(1)}(\frac{i}{\gamma_{n+1}})\sqrt{t}]^q e^{-t^2-2at}dt\Big\}=$$

$$=\frac{1}{2\pi}\sqrt{\frac{2}{\kappa|\varphi''(\frac{i}{\gamma_{n+1}})|}}e^{i(\kappa|\varphi(\frac{i}{\gamma_{n+1}})|+\frac{\pi}{4})}\bar{V}(\frac{i}{\gamma_{n+1}})I_{\text{head}}=\frac{\exp[\kappa|\varphi(\frac{i}{\gamma_{n+1}})|+\frac{\pi}{4}]i}{\sqrt{2\pi\kappa|\varphi''(\frac{i}{\gamma_{n+1}})|}}\mathcal{Y}_w^{\text{head}}(\kappa,\frac{i}{\gamma_{n+1}}), \tag{2.14}$$

where

$$a=\varphi'(\frac{i}{\gamma_{n+1}})\sqrt{\frac{\kappa}{2|\varphi''(\frac{i}{\gamma_{n+1}})|}}\,e^{-\frac{\pi}{4}i},$$

$$\mathcal{Y}_w^{\text{head}}(\kappa,\frac{i}{\gamma_{n+1}})=\frac{1}{\sqrt{\pi}}\bar{V}(\frac{i}{\gamma_{n+1}})I_{\text{head}} \tag{2.15}$$

3. We now consider an arbitrary reflected wave of longitudinal type which is incident on the boundary κ $(\kappa<n)$ at an angle close to the limiting angle, propagates into the next lower medium, and then passes to the surface. In this case the saddle point is located in a neighborhood of the point $\eta=\frac{i}{\gamma_{\kappa+1}}$, which is the branch point of the phase function $\varphi(\eta)$ of maximum modulus. All branch points of the function $V(\eta)$ not contained in $\varphi(\eta)$, may be located either below or above the point $\eta=\frac{i}{\gamma_{\kappa+1}}$. The saddle point η_0 can be located arbitrarily close to the branch point $\eta=\frac{i}{\gamma_{\kappa+1}}$, while $|\eta_0|>\frac{1}{\gamma_{\kappa+1}}$ is always satisfied. The radical $\alpha_{\kappa+1}$ which contains the branch point $\eta=\frac{i}{\gamma_{\kappa+1}}$ will enter both in the terms $V(\kappa,\kappa+1)$, depending on the properties of the layers with numbers κ and $\kappa+1$, and in terms $V(\kappa+1,\kappa+2)$,

depending on properties of the layers $\kappa+1$ and $\kappa+2$. Terms of the form $V(\kappa,\kappa+1)$ (or $V(\kappa+1,\kappa+2)$) consist of products of coefficients determining the properties of reflection and refraction from the boundary with number κ (or $\kappa+1$) both from above and from below. Therefore, the function $V(\eta)$ contains at least eight factors to various powers, which depend on the multiplicity of the wave, containing the branch $\eta=\frac{i}{\gamma_{\kappa+1}}$. The function $V(\eta)$ may then be represented in the form

$$V(\eta)=\bar{V}(\eta)\prod_{q=1}^{8}\left(A_q^{(0)}+\alpha_{\kappa+1}A_q^{(1)}\right)^{\kappa_q}, \tag{2.16}$$

where coefficients with index q = 1,2,3,4 are the complex analogues of the coefficients of reflection and refraction from the κ-th boundary from above and below, while coefficients with index q =5,6,7,8 refer to the same coefficients but from the boundary $\kappa+1$. The phase function is given by

$$\varphi(\eta,t)=tv_s'\eta-\sum_{\ell=1}^{n}{}' m_\ell h_\ell \alpha_\ell - m_{\kappa+1}h_{\kappa+1}\alpha_{\kappa+1}=\varphi_1(\eta)-m_{\kappa+1}h_{\kappa+1}\alpha_{\kappa+1} . \tag{2.17}$$

(In (2.17) the sum with the dash means that the term with number $\ell=\kappa+1$ is missing.) We substitute (2.16) and (2.17) into the Mellin integral (1.2) taken over the principal part of the path of integration and go over to the new complex variable

$$\varkappa=\sqrt{1+\gamma_{\kappa+1}^2\eta^2} . \tag{2.18}$$

We then have

$$\gamma_1=\frac{1}{2\pi i}\int_{(\ell_\varkappa)_+}\bar{V}(\varkappa)\prod_{q=1}^{8}\left[A_q^{(0)}(\varkappa)+\varkappa A_q^{(1)}(\varkappa)\right]\cdot\exp\kappa\left[tv_s'\eta(\varkappa)-h_{\kappa+1}m_{\kappa+1}\varkappa-\sum_{\ell=1}^{n}{}' h_\ell m_\ell\alpha_\ell(\varkappa)\right]\frac{d\eta}{d\varkappa}\,d\varkappa . \tag{2.19}$$

In (2.19) $(\ell_\varkappa)_+$ denotes the principal part of the path of integration in the $\varkappa$ plane, and we have introduced the function

$$\frac{d\eta}{d\varkappa}=\frac{\varkappa}{\gamma_{\kappa+1}\sqrt{\varkappa_0^2+1}} . \tag{2.20}$$

We choose for the functions in (2.19) approximations on the principal segment of the path of integration $(\ell_\varkappa)_+$, which describe the abrupt character of the variation of these functions sufficiently well. Inasmuch as the principal segment of the path of integration in the $\varkappa$ plane can approach arbitrarily close to the point $\varkappa=0$, the variation of the factor $\varkappa$ in $V(\varkappa)$, $\frac{d\eta}{d\varkappa}$, and $\varphi(\varkappa)$ cannot be neglected. We approximate the remaining terms by their expansions in Taylor series in a neighborhood of the saddle point. This is clearly possible if the principal segment of the path of integration is located away from the singularities of these functions.

Thus, the approximate representations which we shall use are the following:

$$V(\varkappa)\approx\bar{V}(\varkappa_0)\prod_{q=1}^{8}\left(A_q^{(0)}(\varkappa_0)+\varkappa A^{(1)}(\varkappa_0)\right)^{\kappa_q},$$

$$\varphi(\varkappa)\approx\varphi_1(\varkappa_0)+\varphi_1'(\varkappa_0)(\varkappa-\varkappa_0)+\frac{\varphi_1''(\varkappa_0)}{2!}(\varkappa-\varkappa_0)^2+h_{\kappa+1}m_{\kappa+1}\varkappa, \tag{2.21}$$

$$\frac{d\eta}{d\varkappa}\approx\frac{\varkappa}{\gamma_{\kappa+1}\sqrt{\varkappa_0^2+1}}$$

Substituting (2.21) into (1.2), replacing the variable of integration $\varkappa$ by t according to the formula

$$\varkappa-\varkappa_0=te^{\frac{3}{4}\pi i}\sqrt{\frac{2}{\kappa|\varphi''(\varkappa_0)|}}, \tag{2.22}$$

and extending the limits of integration to infinity, we obtain the final result in the form

$$y = \frac{1}{\pi\kappa|\varphi''(\xi_0)|}\bar{V}(\xi_0)\frac{\exp[\kappa|\varphi(\xi_0)|+\frac{\pi}{2}]i}{\gamma_{\kappa+1}\sqrt{\xi_0^2+1}}(A_q^{(1)})^{\kappa_q}\cdot\int_{-\infty}^{\infty}(t+a_0)\prod_{q=1}^{8}(t+A_q)^{\kappa_q}e^{-t^2}dt = \frac{\exp[\kappa|\varphi(\xi_0)|+\frac{\pi}{4}]i}{\sqrt{2\pi\kappa|\varphi''(\xi_0)|}}\cdot y_w^{\text{refl}}(\kappa,\xi_0), \quad (2.23)$$

where we have introduced the notation

$$A_q = \frac{A_q^{(0)}+\xi_0 A_q^{(1)}}{A_q^{(1)}}e^{-\frac{3}{4}\pi i}\sqrt{\frac{\kappa|\varphi''(\xi_0)|}{2!}} = \frac{A_q^{(0)}}{A_q^{(1)}}e^{-\frac{3}{4}\pi i}\sqrt{\frac{\kappa|\varphi''(\xi_0)|}{2}}+a_0,$$

$$a_0 = \xi_0 e^{-\frac{3}{4}\pi i}\sqrt{\frac{\kappa|\varphi''(\xi_0)|}{2}}, \quad (2.24)$$

$$y_w^{\text{refl}}(\kappa,\xi_0) = \frac{1}{\sqrt{\pi}}\left(\frac{2}{\kappa|\varphi''(\xi_0)|}\right)^{\frac{1}{2}}\bar{V}(\xi_0)e^{i\frac{\pi}{4}}\frac{1}{\gamma_{\kappa+1}\sqrt{\xi_0^2+1}}(A_q^{(1)})^{\kappa_q}\cdot\int_{-\infty}^{\infty}(t+a_0)\prod_{q=1}^{8}(t+A_q)^{\kappa_q}e^{-t^2}dt.$$

§3. Some Examples and Generalizations to Other Cases

1. Suppose now that

$$\delta_{n+1} < \max(\gamma_\kappa, \delta_\kappa) \qquad (\kappa = 1, \ldots n).$$

On the imaginary axis there are two branch points $\eta = \frac{i}{\gamma_{n+1}}$ and $\eta = \frac{i}{\delta_{n+1}}$ in a neighborhood of which the branch point η_0 may be located. The exact formulas for the principal parts of the displacement fields of the reflected and head waves are derived in the same manner as in subsections 1 and 2 of Section 2. When the saddle point is located in a neighborhood of the point $\eta = \frac{i}{\delta_{n+1}}$ the radical $V(n, n+1)$ must be separated out in the function $\beta_{n+1} = \sqrt{1+\delta_{n+1}^2\eta^2}$ The final formulas in this case are written just as (2.9) and (2.14), but the functions $A^{(0)}$ and $A^{(1)}$ are replaced by $\bar{A}^{(0)}$ and $\bar{A}^{(1)}$, where $V(n, n+1) = \bar{A}^{(0)} + \beta_{n+1}\bar{A}^{(1)}$ and the quantity γ_{n+1} is replaced by δ_{n+1}. We proceed in a similar manner when other types of waves are considered, for example, transverse waves or those admitting changes on the boundaries separating the media. In carrying out the calculations in these cases it may turn out that the saddle point is located alternately in a neighborhood of one, two, or even a greater number of branch points. The basic rules for obtaining expressions for the principal parts of the displacement fields remain the same. The formulas must first be written which define the exact solution of the problem for a wave of the given type. The character of distribution of singularities of the functions in the integrands on the imaginary axis and the position of the saddle point relative to these singularities must be determined. The functions of the integrands are then approximated by some approximate expressions chosen in such a manner as to describe as completely as possible the abrupt variation of these functions near their singularities and also to permit easy computation of the integrals obtained.

The most difficult question is the choice of approximations for the functions of the integrands. This is done in various manners depending on how the applicability conditions for the method are satisfied. To check that these conditions are satisfied analytically is rather complicated. Therefore, it is evidently necessary in certain particular cases to make special investigations of a computational type such as was done, for example, in [5].

2. We shall now illustrate what has been said by obtaining the principal parts of the displacement fields in certain particular cases.

We consider a five-layer elastic system in which the propagation speeds and densities are constant and are given by Table 1. It is clear that the parameters of the given system are related as follows:

$$1 < \frac{1}{\gamma_1} < \frac{1}{\delta_2} < \frac{1}{\delta_4} < \frac{1}{\delta_3} < \frac{1}{\delta_5} < \frac{1}{\gamma_2} < \frac{1}{\gamma_4} < \frac{1}{\gamma_3} < \frac{1}{\gamma_5}.$$

TABLE 1

Layer number	Longitudinal-wave speeds	Transverse-wave speeds	Thickness at given density (m)
I	450	230	60
2	2240	I000	500
3	3900	2070	50
4	3200	I470	50
5	4500	2250	

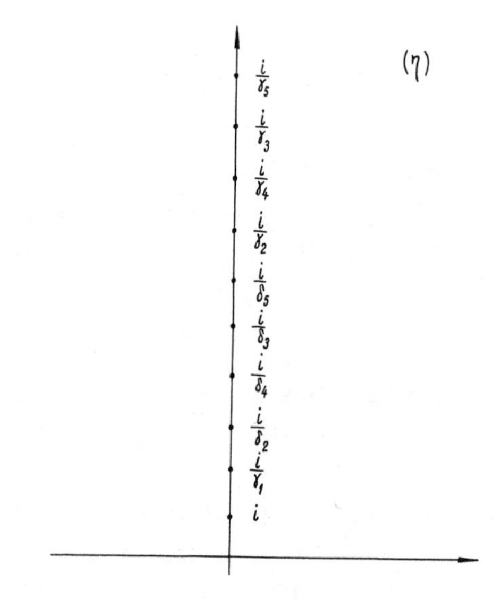

Fig. 2

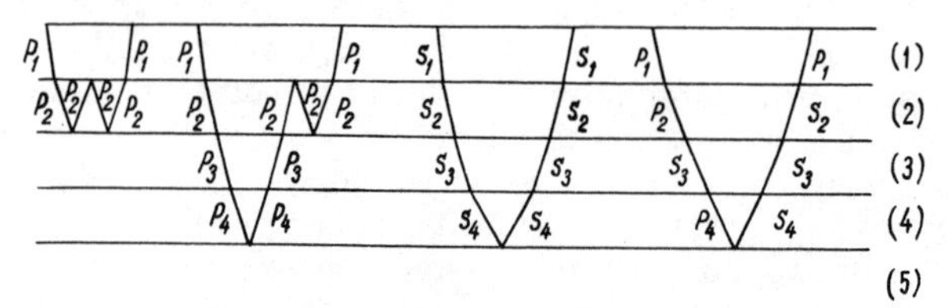

Fig. 3

The corresponding branch points $\eta = \frac{i}{\gamma_\kappa}$ and $\eta = \frac{i}{\delta_\kappa}$ $(\kappa = 1, \ldots 5)$ located on the imaginary axis are shown in Fig. 2.

a) We consider a wave of type $P_1 P_2 P_2 P_2 P_2 P_1$ reflected from the second horizon (Fig. 3). The Mellin integral entering in the exact solution for the vertical component of the displacement is written in the form (1.2) in which the functions $\varphi(\eta, t)$ and $V(\eta)$ are given by

$$\varphi(\eta, t) = t v_s' \eta - 2 h_1 \sqrt{1 + \gamma_1^2 h^2} - 4 h_2 \sqrt{1 + \gamma_2^2 \eta^2},$$

$$V(\eta) = i(1;\eta) P(1,2;\eta) P^2(2,2;\eta) \bar{P}(2,2;\eta) \cdot P(2,1;\eta)\, æ\, (1;\eta),$$

where the coefficients $i(\eta)$, $\varkappa(\eta)$, and $P(\eta)$ denote the complex analogues of the coefficients of directedness of the source, conversion, and refraction and reflection of the longitudinal wave. The distribution of the branch points on the imaginary axis in the present case can be obtained from Figure 2 if branch points corresponding to layers 4 and 5 are removed. The saddle point corresponding to the wave in question will move from $i\infty$ and fall first in a neighborhood of the point $\eta = \frac{i}{\gamma_3}$, and then of the point $\eta = \frac{i}{\gamma_2}$, which is the branch point of maximum modulus of the phase function $\varphi(\eta, t)$. The location of the point η_0 in a neighborhood of $\eta = \frac{i}{\gamma_3}$ corresponds to an observer's location near the point of exit of a head wave of type $P_1P_2P_3P_2P_2P_2P_1$ or $P_1P_2P_2P_2P_3P_2P_1$. The radical $\alpha_3 = \sqrt{1+\gamma_3^2\eta^2}$, which contains the branch $\eta = \frac{i}{\gamma_3}$, enters in the function $p^2(2,2;\eta)$, and the principal part of the wave is obtained directly by applying formulas (2.9) and (2.14) in which $q = 2$ and $n = 2$.

b) We now consider a wave of type $P_1P_2P_3P_4P_4P_3P_2P_2P_1$ (Figure 3). The functions $\varphi(\eta,t)$ and $V(\eta)$ have the form

$$\varphi(\eta,t) = t v_5'\eta - 2h_1\sqrt{1+\gamma_1^2\eta^2} - 4h_2\sqrt{1+\gamma_2^2\eta^2} - 2h_3\sqrt{1+\gamma_3^2\eta^2} - 2h_4\sqrt{1+\gamma_4^2\eta^2},$$

$$V(\eta) = i(1;\eta)P(1,2;h)P(2,3;\eta)P(3,4;\eta)P(4,4;\eta)\cdot P(4,3;\eta)P(3,2;h)\bar{P}(2,2;\eta)P(2,2;\eta)P(2,1;\eta)\varkappa(1;\eta).$$

The distribution of singularities in the η plane is shown in Figure 2. The saddle point does not drop below the branch point $\eta = \frac{i}{\gamma_3}$. If this wave is considered at angles of incidence close to the limiting angle for the second horizon, then it turns out that the saddle point η_0 is located in a neighborhood of the point $\eta = \frac{i}{\gamma_3}$. An expression for the principal part of the wave field is obtained from formula (2.23) in which we put $k_1 = 2$, $k_2 = 1$, $k_3 = 1$, $k_4 = 0$, $k_5 = k_7 = 1$, $k_6 = k_8 = 0$.

3. At the beginning of the section it was remarked that formulas for the principal parts of the displacement fields for transverse waves and waves of interchange type can be obtained in a similar manner. In this regard we consider the following examples.

a) We consider a wave of type $S_1S_2S_3S_4S_4S_3S_2S_1$. The distribution of singularities in the η plane remains the same. The branch point of maximum modulus of the phase function $\varphi(\eta,t)$, which in this case has the form

$$\varphi(\eta,t) = t v_5'\eta - 2h_4\sqrt{1+\eta^2} - 2h_2\sqrt{1+\delta_2^2\eta^2} - 2h_3\sqrt{1+\delta_3^2\eta^2} - 2h_4\sqrt{1+\delta_4^2\eta^2},$$

will be the point $\eta = \frac{i}{\delta_3}$. In moving along the imaginary axis the saddle point η_0 successively passes the branch points $\eta = \frac{i}{\gamma_5}$, $\eta = \frac{i}{\gamma_4}$, $\eta = \frac{i}{\gamma_3}$, $\eta = \frac{i}{\gamma_2}$, $\eta = \frac{i}{\delta_5}$ and arrives at the point $\eta = \frac{i}{\delta_3}$. The function $V(\eta)$ has the form

$$V(\eta) = i(1;\eta)\varkappa(1;\eta)S(1,2;\eta)S(2,3;\eta)S(3,4;\eta)S(4,4;\eta)\;S(4,3;\eta)S(3,2;\eta)S(2,1;\eta),$$

where $i(\eta)$, $\varkappa(\eta)$, and $s(\eta)$ are the complex analogues of the coefficients of directedness of conversion, reflection, and refraction for the transverse wave from the corresponding layers.

We number the points $\frac{1}{\gamma_5}, \frac{1}{\gamma_3}, \frac{1}{\gamma_4}, \frac{1}{\gamma_2}, \frac{1}{\delta_5}, \frac{1}{\delta_3}$ in decreasing order such that $\gamma_5 = n_1$, $\gamma_3 = n_2, \ldots \delta_3 = n_6$. Then in order to obtain expressions for the principal parts of the wave field in a neighborhood of the point $\eta = \frac{i}{n_\nu}$ $(\nu = 1, \ldots 6)$ it is necessary to apply the formulas obtained in subsections 1 and 2 of Section 2 to neighborhoods of the points $\eta = \frac{i}{n_\nu}$, at each stage considering which of the factors of the function $V(\eta)$ contain the branch point $\eta = \frac{i}{n_\nu}$. For example, in order to write down the formulas for a neighborhood of the point $\eta = \frac{i}{n_3}$, it is necessary to choose for the function $\bar{V}(\eta)$

$$\bar{V}(\eta)=i(1;\eta)\varkappa(1;\eta)S(1,2;\eta)S(2,3;\eta)S(3,2;\eta)S(2,1;\eta),$$

represent the product

$$\Pi=S(3,4;\eta)\cdot S(4,4;\eta)\cdot S(4,3;\eta)$$

in the form

$$\Pi=\prod_{K=1}^{3}(A_K^{(1)}+\sqrt{1+\eta^2 n_3^2}\,A_K^{(2)}),$$

and apply the formula of subsection 3 of Section 2.

b) We now consider a wave of interchange type $P_1P_2S_3P_4S_4S_3S_2P_1$ In this case the functions $\varphi(\eta,t)$ and $V(\eta)$ have the form

$$\varphi(\eta,t)=tv_s'\eta-2h_1\sqrt{1+\gamma_1^2\eta^2}-h_2\sqrt{1+\gamma_2^2\eta^2}-h_2\sqrt{1+\delta_1^2\eta^2}-2h_3\sqrt{1+\delta_3^2\eta^2}-h_4\sqrt{1+\gamma_4^2\eta^2}-h_4\sqrt{1+\delta_4^2\eta^2},$$

$$V(\eta)=i(1;\eta)\varkappa(1;\eta)P(1,2;\eta)PS(2,3;\eta)SP(3,4;\eta)\cdot PS(4,4;\eta)S(4,3;\eta)S(3,2;\eta)SP(2,1;\eta).$$

The saddle point η_0 can be located in a neighborhood of the branch points $\eta=\frac{i}{\gamma_5}$, $\eta=\frac{i}{\gamma_3}$, and $\eta=\frac{i}{\gamma_4}$, while $|\eta_0|\geqslant\frac{1}{\gamma_4}$ is always satisfied. In this case also the formulas in which we are interested can be written down by using the results of Section 2.

4. Remarks

a) In order to obtain the displacement fields, it is necessary to be able to evaluate integrals of Fourier-Bessel type in which the integrands are expressions we have calculated. Substituting formulas of type (2.9), (2.14), and (2.23) into the integrals (1.1) and considering integration only over the interval (κ_0,∞), where κ_0 is determined from the condition $\kappa_0 v_s' t \gg 1$, (which is equivalent to considering the sharply varying parts of the wave field), it is possible to obtain final computational formulas for the components of the displacement fields. If the expressions for the principal parts of (1.2) corresponding to reflected and head waves are written in the form

$$S^{\text{refl}}(\kappa,t)=\frac{\exp[\kappa|\varphi(\varkappa_0,t)|+\frac{\pi}{4}]i}{\sqrt{2\pi\kappa|\varphi''(\varkappa_0,t)|}}\,y_q^{\text{refl}}(\kappa,\varkappa_0),$$

$$R^{\text{refl}}(\kappa,t)=\frac{\exp[\kappa|\varphi(\varkappa_0,t)|+\frac{\pi}{4}]i}{\sqrt{2\pi\kappa|\varphi''(\varkappa_0,t)|}}\,y_w^{\text{refl}}(\kappa,\varkappa_0),\quad(\varkappa_0=\eta_0,\tau_0)\tag{3.1}$$

$$S^{\text{head}}(\kappa,t)=\frac{\exp[\kappa|\varphi(\frac{i}{\gamma_{n+1}})|+\frac{\pi}{4}]i}{\sqrt{2\pi\kappa|\varphi''(\frac{i}{\gamma_{n+1}})|}}\,y_q^{\text{head}}(\kappa,\frac{i}{\gamma_{n+1}}),$$

$$R^{\text{head}}(\kappa,t)=\frac{\exp[\kappa|\varphi(\frac{i}{\gamma_{n+1}})|+\frac{\pi}{4}]i}{\sqrt{2\pi\kappa|\varphi''(\frac{i}{\gamma_{n+1}})|}}\,y_w^{\text{head}}(\kappa,\frac{i}{\gamma_{n+1}}),\tag{3.2}$$

then it is found that the components of the displacement can be represented by the formulas

$$q=\frac{-1}{\pi\sqrt{\tau|\varphi''|}}\Big[\int_{\nu_0}^{\infty}(\mathrm{Re}\,y_q\frac{a_\nu(T)}{\nu}-\mathrm{Im}\,y_q\frac{b_\nu(T)}{\nu})\sin\nu\tau\,d\nu+\int_{\nu_0}^{\infty}(\mathrm{Re}\,y_q\frac{b_\nu(T)}{\nu}+\mathrm{Im}\,y_q\frac{a_\nu(T)}{\nu})\cos\nu\tau\,d\nu\Big]$$

$$W=\frac{-1}{\pi\sqrt{\tau|\varphi''|}}\Big[\int_{\nu_0}^{\infty}(\mathrm{Re}\,y_w\frac{b_\nu(T)}{\nu}+\mathrm{Im}\,y_w\frac{a_\nu(T)}{\nu})\sin\nu\tau\,d\nu+\int_{\nu_0}^{\infty}(\mathrm{Im}\,y_w\frac{b_\nu(T)}{\nu}-\mathrm{Re}\,y_w\frac{a_\nu(T)}{\nu})\cos\nu\tau\,d\nu\Big].\tag{3.3}$$

In these formulas we have introduced the notation

$$\varphi = \begin{cases} \varphi(x_0, t), \quad (x_0 = \eta_0, \varepsilon_0), \text{ for reflected waves,} \\ \varphi\left(\frac{i}{\gamma_{n+1}}, t\right) \text{ for head waves.} \end{cases}$$

$$\gamma_q = \gamma_q^{\text{refl}} \quad \text{or} \quad \gamma_q^{\text{head}},$$

$$\gamma_w = \gamma_w^{\text{refl}} \quad \text{or} \quad \gamma_w^{\text{head}},$$

$$v = \begin{cases} \kappa v_s' |\eta_0| \text{ for reflected waves,} \\ \kappa \frac{v_s'}{\gamma_{n+1}} \text{ for head waves.} \end{cases} \tag{3.4}$$

$$a_\nu(T) = \int_0^T f'(u) \sin \nu u \, du,$$

$$b_\nu(T) \int_0^T f'(u) \cos \nu u \, du,$$

in which the function $f(t)$ defines the time dependence of the source and is different from zero in the interval $(0, T)$.

b) In the integrals of formulas (2.9), (2.14), and (2.23) for the principal parts, which we obtained in subsections 1, 2, and 3 of Section 2, parameters enter which depend on the distance from a particular point, on the elastic properties of the medium, and also on the value of κ (or ν).

In analogy to simpler cases, for large values of these parameters the original formulas go over into formulas obtained by the usual method of stationary phase.

c) In carrying out computations it is sometimes expedient to approximate the functions $V(\eta)$ (or $U(\eta)$) by expressions of the form

$$V(\eta) \approx \bar{V}(\eta) \left[\frac{A^{(0)}(\eta_0) + \alpha_{n+1} A^{(1)}(\eta_0)}{B^{(0)}(\eta_0) + \alpha_{n+1} B^{(1)}(\eta_0)} \right]^q. \tag{3.5}$$

For approximations of the form (3.5) it is easy to obtain the corresponding formulas for the principal parts by repeating exactly the derivation of the formulas of Section 2.

LITERATURE CITED

1. Petrashen', G. I., "The general quantitative theory of reflected and head waves in layered media with plane-parallel separating boundaries," in: Problems in the Dynamical Theory of Propagation of Seismic Waves [in Russian], No. 1, Gostoptekhizdat (1957).
2. Smirnova, N. S., "Calculation of the wave fields in a neighborhood of singular points," in: Problems in the Dynamical Theory of Propagation of Seismic Waves [in Russian], No. 6, Gostoptekhizdat (1962).
3. Petrashen', G. I., "Propagation of elastic waves in isotropic layered media," Uch. Zap. Len. Gos. Univ., No. 162 (1952).
4. Smirnov, V. I., A Course in Higher Mathematics [in Russian], Vol. 3, Ch. 2, Leningrad (1949).
5. Smirnova, N. S., Abstract of Candidate's Dissertion [in Russian], Izd. Len. Gos. Univ. (1963).